Physics of Δc Mechanics

A Feynman Path Action Approach to Particle Dynamics

DT Froedge

Formerly Auburn University

dt.froedge903@topper.wku.edu

V031616

Hardcover ISBN: 979-8-218-34717-8
Paperback ISBN: 979-8-218-38312-1
Ebook ISBN: 979-8-218-36658-2

Cover design, book design, and typesetting by Mayfly Design

Library of Congress Catalog Number: 2024900162

First Printing: 2024
Printed in the United States of America

Contents

The version number at the start of all papers (VDDMMYY)
is the last date the paper was edited

Preface

This is the presentation of a field-free, particle based theory of mass and photon interaction, the same as envisioned by Wheeler and Feynman in the 1940's. The difference in their view was that the elementary particles were massive whereas the particles at the core of this development are photons. By making assumptions about the character and properties of photons' interaction, the separate fields identified as electric, gravitation, and strong can be duplicated without any reference to "charge".

The basic concept is that the probability density of moving Feynman photons from one mass particle changes the direction and velocity of Feynman photons from another. This change in the index of refraction thus moves the direction and velocity of the particles and styles the theory as: Δc Mechanics.

The purpose of this book is not as an advocacy of the presented theory, but to preserve the development findings as they came about, and as a guide for someone later. The author is getting old and does not have the acuity of earlier years, and thus the first paper in this book is the last.

Readers will find the older papers have both mathematical and conceptual errors, later discovered to be wrong. Those papers should have, but have not, been revised, and as the project has further developed, the results have shown the developed concept to be "too right, to be wrong".

DT Froedge
October 11, 2023

DT Froedge

Academics

B.S. Physics, Mathematics. Western. Kentucky University, 1965
M.S Physics. University of Tennessee, 1967
PhD, Physics, Auburn University, Academic Course Completed, 1969

Professional

American Physical Society, 2005-Present
License Professional Engineer, PA, CT, NC TX, AL.
CEO, GeoSonics & Vibra-Tech Engineers Inc.
E-mail: DT.Froedge903@topper.wku.edu

Introduction:
Δc Mechanics

Δc Mechanics is an alternate approach to the physical interaction between mass and photons by way of the Feynman action path probabilities, offering a new approach to mechanical dynamics and particle structure. [1-10]. It is based on the Feynman view of QED that photons going from one point to another follow all paths and there is a probability that the particles actually exist on those paths. Additionally photon densities moving in one direction retard opposite moving photons and thus create a change in c, Δc, as a result of their presence.

In the 1940's, John Wheeler and Richard Feynman developed a vision of a field-free particle based theory of physics. The Wheeler-Feynman absorber theory developed from this with the view that physical interactions between a source and an absorber were particle based. This field-free idea came from Feynman, with Wheeler as the primary developer. The theory was based on Feynman's ideas and Wheeler's insight and experience. Feynman's dissertation was based on the theory but it had problems and was eventually abandoned.

Feynman continued the pursuit however, and while attempting to put the theory into a quantum basis, developed the idea of summing probability weighted quantum paths. Wheeler titled this idea as the "sum over all history" method of quantum electrodynamics, and for the work Feynman received the Nobel Prize.

Δc Mechanics starts with Feynman's action paths and continues a particle based theory. It extends Feynman's presumption in that there is an actual probability of the Feynman path photons being on these paths. This is evidenced by measurements of the Anomalous Magnetic Moment and the Aharonov-Bohm effect. It is presumed that photons taking the action path not only have delays, but the probability density of these photons being there affects the speed of light and the motions of other particles.

Δc Mechanics is not based on electron and positron as the primary particles as Wheeler had envisioned, but based on photons and photon-photon interactions. Electrons are in fact a composition of two photons bound together by their own self interaction, and it is presumed that all mass is composed of the three primary leptons: the electron, the muon, and the tauon photons. Each of these three leptons is composed of the binding of two specific conjugate photons. The photons of these particles have wavefunctions and satisfy Dirac type particle functions. Only photons have wavefunctions [1].

The concept of charge as a substance embedded in mass that provides the mechanism for attraction and repulsion is absent in Δc Mechanics. The attraction and repulsion is related to the physical mechanics of the interaction. The rotating bound Planck size photons that create the electron and other particles provide the mechanisms for all particle forces.

Order of Developments
of Δc Mechanics

V101023

After a review of Feynman's sum over all action paths of photons moving from one point to another, and presuming that there was actually the probability of the photon being there, consideration was given to the possibility that an interloping photon would be altered by the probability.

Presuming the photon is a rotating Planck particle inside the Compton radius of the concept developed that a Planck size particle passing thru the Compton radius of another photon has a probability of intersection and thus collectively probability of a slowdown of a probability and a change in c.

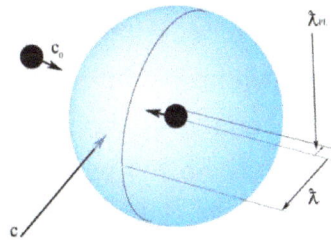

$$\frac{\Delta c}{c_0} = \frac{\sigma_{PL}}{\sigma_e} = \frac{\lambda_{PL}}{\lambda_e}$$

The first consideration was gravitation, and Schapiro's value of the change in c from a gravitational mass, [2], [3], the relation between delta c and mass could be developed.

$$\frac{\Delta c}{c_0} = \frac{2Gm}{r} \qquad \Delta c = c_0 - c \qquad (1)$$

From this then the probability density of Feynman photons as a function of the radius could be originated. $\Delta c / \Delta d$.

1

An increase in the oncoming density of photons, decrease the velocity of light thus the density of photons necessary to establish the current velocity of light can be calculated. By using current estimates of the number of protons in the universe and calculating the density of the Feynman photons, it turns out to match the current velocity. This density can be viewed as the vacuum polarization or background density for Feynman photons in the universe that sets the velocity of light. In any direction a photon moves it encounters an oncoming flow probability density of about 10^39 photons/cm/ sec.[8]

Gravitation is a spherically symmetric distribution of Feynman photons emanating from the internal path actions of mass particles in nuclei that are not spin aligned. There is no spin stabilization and thus the paths as well as the photon probability paths are random.

Charge Effects

The next consideration was the prospect of two photons revolving together encountering the probability density of the other. As they rotate, they are spin stabilized thus he probability density moves in one direction that lies in a plane.

Each time they revolve increases the oncoming photon density at a point, thus slowing down probability density of photons moving opposite. Slowing down is equivalent to increasing the index of refraction in the proximity of the other particle and thus bending the trajectory and pulling the particles together. At the right frequency of rotation for the Planck particles the photons can bind and create the electron at the radius, $\Re_0$.[8]

$$\Re_0 = \sqrt{2} \lambda_{PL} v_e \tag{2}$$

$\Re_0$. is the Electron Creation Radius, and is the binding radius of two Planck photons that forms the electron, λ_{PL} is the Planck particle radius, and v_e is a unitless number equal to the Compton frequency of the electron. $v_e = h / m_e c_0^2$. $\Re_0$ is the radius at which the repeating encounter density experienced by rotating photons matches the background encounter density of Feynman photons in the universe, v_e^2.

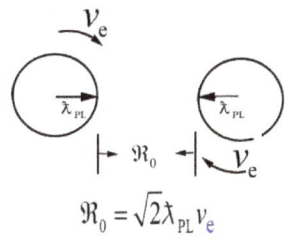

$$\mathfrak{R}_0 = \sqrt{2}\lambda_{PL}v_e$$

Electron Creation Radius

Photons have polarization, with right and left handed versions, thus photons with an opposite polarization can also bind creating the positron. The frequency of the electron is about 10^{20} cycles per second and the encounter probability density is the product of the frequencies ~ 10^{40} which approximately matches the background density of ambient photons, and bends the photon into a circle, $\mathfrak{R}_0$. Eq.(2)

Gravitation is generated by a random direction of the ambient photon emanating from a mass, but electrons are generated by a pair of spin aligned particles. Electric effects are generated by planar rotating photons and the interacting particles lie in a plane.

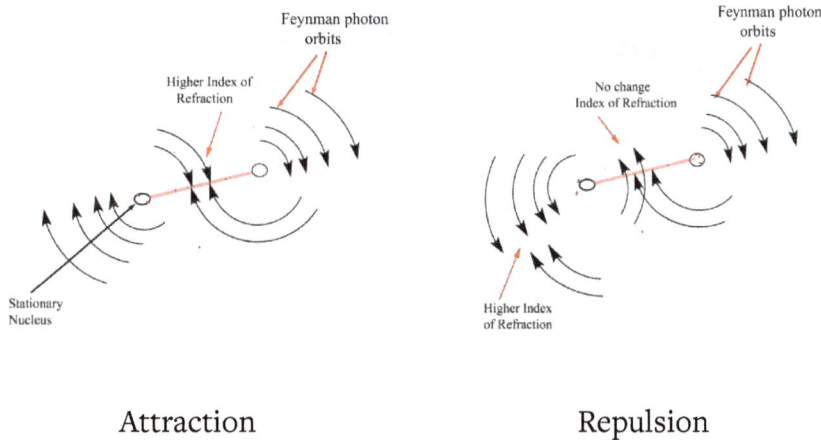

Attraction   Repulsion

Fig.4 Attraction and repulsion by photon density

Electrons with opposite spin alignment engage in a planar motion and can create both attractive and repulsive forces that have the effect of charge. The concept of charge is discovered not to be a substance distributed in a mass, but the directional interaction of the

3

spin aligned opposite polarized Feynman photon probability density. Photons moving in the same direction have no effect on each other whereas photons moving opposite directions decrease their mutual velocity

Interacting Electron Positron
Rotation Disks

Fig. 5

The ratio of the planar photon interaction of the ½ spin electron-positron and the random spherical photon interaction of gravitation is about v_e^2, or 10^40 orders of magnitude.

Fine Structure Constant

The anomalous g factor $g_A = (g/2-1)$, increases the radius of the electron due to the Feynman path integrals that delays the electrons orbit, and has been has been incorporated into the to the value of the fine structure constant

Ground States

Once the value of the fine structure constant is found [6], then the atomic levels are just n values of the Rydberg energy.

$$R_\infty = \frac{1}{2}\left[\frac{\left(\sqrt{2}\,\lambda_{PL}v_e\right)}{n\,\lambda_e\,g_A^2}\right]^2 m_e c_0^2 \quad = \quad \frac{1}{2}\frac{\alpha^2}{n^2}\,m_e c_0^2 \quad \rightarrow 13.6\,ev/n^2 \quad (3)$$

The ground state of the atomic energy level at n = 1 is:

$$E_0 = \frac{1}{2}\left[\frac{\left(\sqrt{2}\,\lambda_{PL}v_e\right)}{\lambda_e}\right]^2 = \frac{1}{2}\left(\frac{\mathfrak{R}_0}{\lambda_e}\right)^2 \qquad (4)$$

It is also the ground state energy for nuclear particles. The energy level is slightly lower than the Rydberg levels Eq. (3), by a factor of g_A^2

The integral of the potential energy from a point r to ∞ is the escape energy, which will apply for both atomic and nuclear particles. The escape energy for both atomic and nuclear particles is then

$$\varepsilon = \int_r^\infty \frac{1}{2}\left(\frac{\mathfrak{R}_0}{r}\right)^2 = \frac{1}{2}\left(\frac{\mathfrak{R}_0}{r}\right) \qquad (5)$$

The separation between positive and deficit kinetic energy is the ground state kinetic mass for atomic as well as nuclear particles at n = 1 is:

$$\varepsilon_0 = \frac{1}{2}\left(\frac{\mathfrak{R}_0}{\lambda_e}\right) \qquad (6)$$

The electron is the ground state particle of the atomic interaction thus the ratio of the atomic levels to the electron is an integral constant α^2/n^2. It was a serendipitous finding that the ratio of nuclear particle masses to the electron mass had similar values of the kinetic mass and the binding mass that were exactly equal to states of other nuclear particles.

A series of these relations were found and developed in " Nuclear Particle Structure in Δc Mechanics". Included below are some examples.

Tau Quark Identity

$$\frac{1}{2}\left(\frac{\lambda_\tau}{\lambda_e}\frac{\lambda_\tau}{\lambda_e}\right)=\frac{1}{2}\left(\frac{2^3\lambda_P}{\lambda_e}\right)\left(\frac{2^3\lambda_P}{\lambda_e}\right)\left(\frac{2^3\lambda_P}{\lambda_e}\right)$$

Binding 2 Tauons = Binding 3 Quarks

The state energy level of two bound tauons is equal to the state energy level of the three bound quarks

The mass of the quarks are referenced here to the proton, however the bound quarks have a different fine structure component of mass.

Another Identify that has been found is the relation between the Tauon and the Z boson.

$$\frac{\frac{1}{2}\left(\frac{\lambda_\tau}{\lambda_e}\right)^2}{\frac{1}{2}\left(\frac{\mathfrak{R}_0}{\lambda_e}\right)^2}=\frac{\left(\frac{\lambda_Z}{\lambda_e}\right)}{\frac{1}{2}\left(\frac{\mathfrak{R}_0}{\lambda_e}\right)} \qquad m_Z=\frac{176871.72 \ \text{electrons}}{90.3833\text{Gev}}$$

This relation shows the state value of two bound tauons referenced to the ground state. E_0, is equal to the kinetic mass of the Z boson referenced to the kinetic mass ground state ε_0.

Physics of Δc Mechanics

V101023

This paper is planned to be the introduction to a compiled edition of all the referenced paper in the development of Δc Mechanics.

The development of the theory is far from complete, but enough of the results are present to show that it is an important approach to the physics of mass creation and interaction.

The electron, an $SU(2)$ particle allows analytical solutions of the atomic state levels, and the identities have defined a number of states and relations between states in the nucleus, but to define the nuclear states analytically will require the development of the muon and tauon wavefunctions. From what we know of the Standard Model analytical development of the wavefunction of those particles will be in the Lee algebra of $SU(2)SO(3)$.

Index

Δc Basis

Photons

The photon is postulated to be a Planck particle rotating at the Compton frequency surrounded by a probability flow of position. The wavefunction defines the probability flow in the photons location, the electric vector being the flow and the B vector being the return flow. There are two polarizations right and left handed designating the relation between the E, B and the Poynting vectors. The positive electron has two photons with one polarization and the negative electron (positron) has two photons with the opposite polarization.[4]

In one revolution the probability flow moving radially at c extends primarily to the first Compton radius. As the frequency, and energy, increases, the Compton radius the Compton radius decreases. The probability density of the flow exists to a considerable radius beyond

Compton radius but the photon is only the size of the bare Planck particle.

The probability density flow that extends further into space, and is defined by the Feynman action paths version of QED, that extends over all space. As the particle rotates the probability density also rotates, and encounters the probability flow of other photons. The primary photon for the electron can mathematically be defined by the four space vectors of geometric algebra.

The Electron

The electron is composed of two special photons, that revolve in orbit with each other at rotational frequency of half the electron Compton frequency, $v_e = c_0 / 2\pi \lambda_e = m_e c_0^2 / \hbar$, The flow probabilities are opposite and thus retard the flow probability of the other. This reduction in flow velocity is a reduction in c and thus an increase in the index of refraction. As the particles come closer the density goes up and the velocity goes down creating a radial gradient in c that is sufficient to bind the particles together.

From the paper on vacuum polarization [8] it is found that the radius of the bound photons is related to the background density of Feynman photons in the universe and is designated as the Electron Creation Radius, $\mathfrak{R}_0 = \sqrt{2} \lambda_{PL} v_e$. It is the product of the radius of the Planck particle radius and the electron Compton frequency. This radius results from the ratio of the background density the Feynman Photon to the increased density of the photon experienced by the rotationally increased density of the other particle

Each revolution increases the oncoming photon density at a point, thus slowing down probability density of photons moving opposite. Slowing down is equivalent to increasing the index of refraction in the proximity of the other particle and thus bending the trajectory and pulling the particles together. At the right frequency of rotation for the Planck particles the photons can bind and create the electron at the radius, $\mathfrak{R}_0$.[4]

$$\mathfrak{R}_0 = \sqrt{2}\lambda_{PL}\nu_e \qquad (1)$$

$\mathfrak{R}_0$. is the Electron Creation Radius, and is the binding radius of two Planck photons that forms the electron, λ_{PL} is the Planck particle radius, and ν_e is a unitless number equal to the Compton frequency of the electron. $\nu_e = h / m_e c_0^2$. $\mathfrak{R}_0$ is the radius at which the repeating encounter density experienced by rotating photons matches the background encounter density of Feynman photons in the universe, ν_e^2.

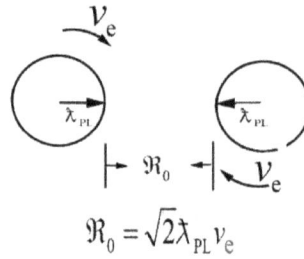

Fig. 1 Electron Creation Radius

Primary Leptons

There are three primary leptons; the Electron, Muon and Tauon, and each of these have two internal photons, and each have their own photon types having wavefunctions, namely the Electron, Muon and Tauon photons. Each of these three leptons consists of a pair of the specific photons having wavefunctions with integral dimensional solutions for opposite moving conjugate pairs. The two electron photon wavefunction can be specified in SU (2) geometric algebra. The Muon and Tauon wavefunctions, yet to be found, must be specified in a combination of SU (2) SO(3), consistent with the standard model.

Only the photons have wavefunctions. The composite pairs have Dirac type solutions illustrated in [1] which shows the rest mass generated from the sum of the four-differential Wavefunctions.

The Electron Mass

Rest mass is the locally confining of photons, and the mass ground state particle is the electron, being the composition of two self-confined photons. The primary leptons are the primary constituents of particle mass, and all other particles are combinational solutions and bindings of the three leptons.

The free electron frequency on the CTR here v_e is not electron frequency, but a unitless ratio representing the number of cycles the photon makes to self-encounter the same photon probability flow density n/ sec/ cm² in free space[8]. It is numerically it is equivalent to v_e

Bound Particles

Electrons are composed of two orbiting photons, thus the probability density of the Feynman photons lie in a plane. Opposite particles having opposite aligned spin turn in the same direction thus in the space between the particles the interacting photon-photon collision probability densities are opposite thus reduce c or increase the index of refraction. This reduction in the index of refraction pulls the particles together as well as aligning the planes and spin axis [8], "opposite" electrons with rotating Feynman photons engage, the B vectors anti-align (Fig. 2 A).

When two particles are alike (both positive and both negative), the B vector's anti-alignment puts the interior photon flow in the same direction, Photon flow probability in the same direction does not interact thus for probability flow density between the particles there is no attraction. There is however a decrease in the index of refraction exterior to the electrons orbit and index of refraction pulls the particles apart. (Repulsion)

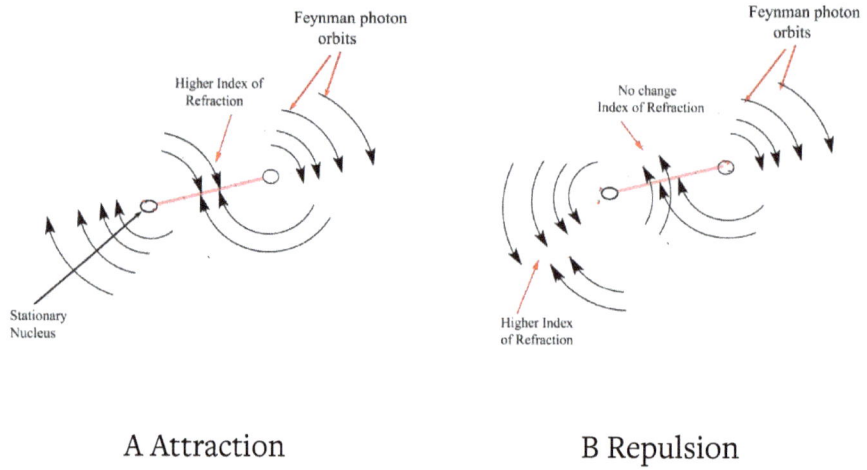

A Attraction   B Repulsion

Fig. 2,

This is the mechanics of repulsion and attraction heretofore ascribed to charge. [8].

Specifics for atomic interaction

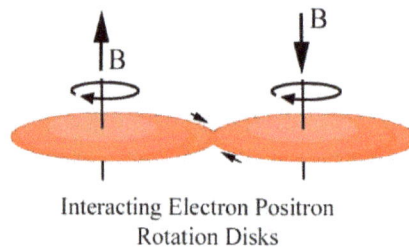

Interacting Electron Positron
Rotation Disks

Electrons and positrons, composition of orbiting polarized photons interact by engaging their respective opposite going probability density.

12

For an electron (Eq.(9)), the probability density for photons in the plane of rotation is:

$$P = \frac{\lambda_{PL} \nu_e}{r} \qquad (2)$$

The expression for electric interaction equivalent to Eq.(10), but with the repetition rate of the aligned disks ν_e, and is:

$$\frac{\Delta c}{c_0} = \frac{1}{2}\left(\frac{\sqrt{2}\lambda_{PL}\nu_e}{r_1}\right)\left(\frac{\sqrt{2}\lambda_{PL}\nu_e}{r_2}\right) \qquad (3)$$

The 1/2 difference comes from the convention shown next that the Fine Structure Constant has a $\sqrt{2}$ included that the gravitational constant does not.

The term in parenthesis is better recognized as:

$$\frac{\Delta c}{c_0} = \frac{\sqrt{2}\lambda_{PL}\nu_e}{r_1} = \left(\frac{\sqrt{2}\lambda_{PL}\nu_e}{\lambda_e}\right)\frac{\lambda_e}{r} \qquad (4)$$

In which the angular momentum $r = n\lambda_e$ as noted for particles in gravitation in Eq.(9), The term in parenthesis is conventionally designated as the Fine Structure Constant, the accuracy of which discussed in another paper.*

$$\alpha = \frac{\sqrt{2}\lambda_{PL}\nu_e}{\lambda_e g_A} \qquad (5)$$

Eq.(3), can now be written as

$$\frac{\Delta c}{c_0} = \frac{\Delta\varepsilon}{m_e c_0^2} = \frac{1}{2}\frac{\alpha^2}{n^2} \qquad (6)$$

This is the Rydberg energy levels of the hydrogen atom.

$$\frac{\Delta\varepsilon}{\varepsilon_0} = \frac{1}{2}\left(\frac{\mathscr{R}_0}{n\lambda_e}\frac{\mathscr{R}_0}{n\lambda_e}\right) \rightarrow \Delta\varepsilon \approx \frac{1}{2}\frac{\alpha^2}{n^2}mc_0^2 \qquad (7)$$

Note the approximation to the Rydberg energy term in Eq. (7). There is a slight difference due to the anomalous gyromagnetic ratio (see Appendix II for explanation)

The Basic Gravitation and Electric Feynman photon Interaction

The Δc interaction of the Feynman photons inducing a change in c thus creating an energy Potential via $\varepsilon = mc_0^2$, is the same for both Electric and gravitational. It should be noted that the induced potential is not an electric or gravitational potential but an energy potential.

Δc Gravitation

The Feynman photon probability densities from the internal action paths of all particles exist throughout space, but because of the randomness of the internal paths and spin they have a random probability of direction. At a point in space, the Feynman photons generated by a non-aligned particle does not have a repetitious direction, and the change in the speed of a random photon passing is only changed by the photons moving in an opposite direction.

For gravitation the equivalent Feynman photon-photon interaction probability density is the product of the nonaligned random individual probability density.

The probability density of a Feynman photon at a distance r from a photon is has been found to be:

$$ P = \left(\frac{\lambda_{PL}}{r} \right) \quad \rightarrow \quad P = \left(\frac{\lambda_{PL}}{\lambda} \frac{\lambda}{r} \right) \tag{8} $$

λ_{PL}, is the Planck particle radius and D, is the Compton radius of the photon $\lambda = h / mc$.

When two particles engage as a bound pair or, a system, Dirac & Schrodinger solutions are possible for systems when the Compton radii have integral values, or when the angular momentum has an integral value of the Planck constant, h. $r = n\lambda$ thus the probability density for a particles that are solutions has:

$$P = \frac{\lambda_{PL}}{\lambda} \frac{1}{n} \tag{9}$$

If this is the probability density of a single particle, then the probability density of the collision or encounter of photons from the particles would have to be:

$$\frac{\Delta c}{c_0} = 2P_1 P_1 = 2 \frac{\lambda_{PL}}{\lambda_1} \frac{\lambda_{PL}}{\lambda_2} = \frac{2\lambda_{Pl}^2}{\lambda^2} \tag{10}$$

The 2 in the relation account for the finite radius of two encountering photons.

Discussed elsewhere is the relation of Δc being proportional the probability of intersection with an opposite going photon, and the vacuum photon density of free space [8].

Noting that, $\lambda = h / mc$ and the gravitational constant in fundamental elements is $G = \lambda_{PL}^2 c^3 / \hbar$. From Eq. (10), the gravitational relation between G and the velocity of light is:

$$\frac{\Delta c}{c_0} = \frac{2Gm_1}{c^2 r} = \frac{2\mu}{r} \tag{11}$$

This can be summed over all the masses and provides the Schapiro time delay measured as measured by light passing the sun.

It is easy to see that this relation provides the proper gravitational force on a test mass m_2 in the presence of m_1.

$$\frac{\Delta c}{c_0} = \frac{m_2}{m_2} \frac{2Gm_1}{c^2 r} \rightarrow \frac{m_2 c^2 \Delta c}{c_0} \rightarrow \Delta(m_2 c^2) = \frac{2Gm_1 m_2}{r} \tag{12}$$

Or:

$$\Delta(m_2 c^2) = \frac{2Gm_1 m_2}{r} \tag{13}$$

The left term is the change in the total relativistic energy and the left is the energy potential. The derivative with respect to r $d(m_2 c^2)$, gives the force on a particle falling into a gravitational field.

This is a non-dimensional relation for the specific energy change, in the total relativistic energy of the m_2 particle as a function of the distance r from m_1. There is a reciprocal relation for the action of the other particle on the first, but for gravitation the reciprocal effects are small, [10].

The Differences Between Gravitation and Electrical Forces

There is very little difference in the mechanism of electric and gravitational interaction. The notable difference is that the Feynman photons generated by composite particles having no net spin such as the atom and neutron, are generated by random non-aligned internal paths. That makes the Feynman photons from the mass at any given location having a random direction. There is no inertial directional stability, thus the photon paths do not have directional stability. The density in any position is proportional to 1/r, but the direction is random creating a non-directional increase in the density of photons in space and a general decrease in c.

The Feynman photons of the electrons, on the other hand, are generated by a pair of bound rotating photons having a stable axis of rotation. The probability density at a point in space is in a rotating plane, and the photon density repeats at the frequency of the electron's rotation v_e. When the, +/- electrons interact their rotation planes line up and the interaction appears to be planar symmetric

This planar rotation creates planar rotating photons that encounter other electron increasing the opposite photon density by a factor of, $v_e^2 \approx 10^{40}$. This is the physical cause creating the huge differences in the force of electricity and gravitation. Without the increase in the photon density as the result of the rotation, the attraction of electrons is exactly that of gravitation.

For electrons bound in a helium atom, the planes of the two electron rotation planes are perpendicular. The multiple planes in heavier particles move the subsequent shells outward defining the mechanism of the Pauli Exclusion Principle.

Atomic and Nuclear Interaction

Atomic

In previous papers [1-9], the Feynman path integral probability for the relation between the change in c and the energies energy of two opposite electrons has been expressed as a change in the binding potential energy generated by the interaction of two binding particles, as:

$$\frac{\Delta c}{c_0} = \frac{m_B}{m_e} = \frac{1}{2}\left(\frac{\mathfrak{R}_0}{r_1}\frac{\mathfrak{R}_0}{r_2}\right) = \frac{m_e}{m_e} - \frac{m_K}{m_e} \tag{14}$$

The last two terms are the difference in the mass of the free electron and the kinetic inertial mass of one of the bound electrons.

For an atomic system the binding energy solutions occur when values of $\mathfrak{R}_0 / n\lambda_e = \alpha$:

$$\frac{m_B}{m_e} = \frac{1}{2}\left(\frac{\mathfrak{R}_0}{r_1}\frac{\mathfrak{R}_0}{r_2}\right) \rightarrow \frac{1}{2}\left(\frac{\mathfrak{R}_0}{\lambda_e}\frac{\mathfrak{R}_0}{\lambda_e}\right) \rightarrow \frac{1}{2}\frac{\alpha^2}{n^2} \tag{15}$$

(See Appendix II for details)

Thus the Hamiltonian solution for an atomic particle is:

$$\frac{m_e}{m_e} = \frac{1}{2}\frac{\alpha^2}{n^2} + \frac{m_K}{m_e} \tag{16}$$

For a particle bound in a system the terms in parenthesis in Eq.,(15) and the fine structure term in, Eq.(16), represents the binding energy or the deficit energy when the photons are radiated away. The kinetic mass is then the residual mass of the electron. The energy of the absent photons still exists somewhere, but is not part of the inertial mass.

From earlier developments [4], in Eq.(14), $\mathfrak{R}_0$ is the Electron Creation Radius, $\mathfrak{R}_0 = \sqrt{2}\lambda_{PL}v_e$. It is the binding radius of two Planck photons when they form the electron, λ_{PL} is the Planck particle radius, and v_e is a unitless number equal to the Compton frequency of the electron. $v_e = h / m_e c_0^2$

From the paper on vacuum polarization [4], the electron is the first Lorentz rest mass level in the universe and the ground state level for the mass of all other particles. The ratio of the electron, rest energy to the rest energy of other mass particles is the ratio of the particles energy to its ground state energy.

Into the Nucleus

The binding energy is based on the probability density of Feynman photons rotating around the center of mass and there is no discontinuity if the distance r is greater or less than $\mathfrak{R}_0$. In Eq.(14), if r becomes less than $\mathfrak{R}_0$, $r < \mathfrak{R}_0$, the kinetic mass becomes negative. The equation cannot be valid unless kinetic energy is injected into the system, or the binding energy is less than the kinetic mass energy.

$$\frac{1}{2}\left(\frac{\mathfrak{R}_0}{r_1}\frac{\mathfrak{R}_0}{r_2}\right) - \frac{m_K}{m_e} = \frac{m_e}{m_e} \tag{17}$$

Once energy is injected into the system, the Lagrangian can have integral solutions, but when the particle acquires a state, the same as in the case for atomic particles, the binding energy is radiated away. The particle is then left as a deficit particle, or in an energy hole.

$$\frac{1}{2}\left(\frac{\mathfrak{R}_0}{n\lambda_X}\frac{\mathfrak{R}_0}{n\lambda_X}\right) - \frac{m_K}{m_e} = \frac{m_e}{m_e} \tag{18}$$

The difference between the atomic particle and the nuclear particle is that for the atomic particle the deficit is less than the residual kinetic mass of the electron, whereas for the nuclear particle the deficit energy exceeds the kinetic mass energy of the residual electrons.

In the nucleus, a particle that has acquired a state that is revolving around another particle retains some kinetic energy. That energy is not radiated away as state binding energy and the particle retains some mass in addition to the elections. The Lagrangian expressed in Eq.(18), at integral values of $\hbar$ has state solutions that define the nuclear mass particles. States showing this relation are shown in the enclosed identities of [10].

A graph showing the atomic and nuclear mass levels and the ground level values is presented in Fig.3. This illustrates the continuation of the Δc binding function from atomic into nuclear states [10].

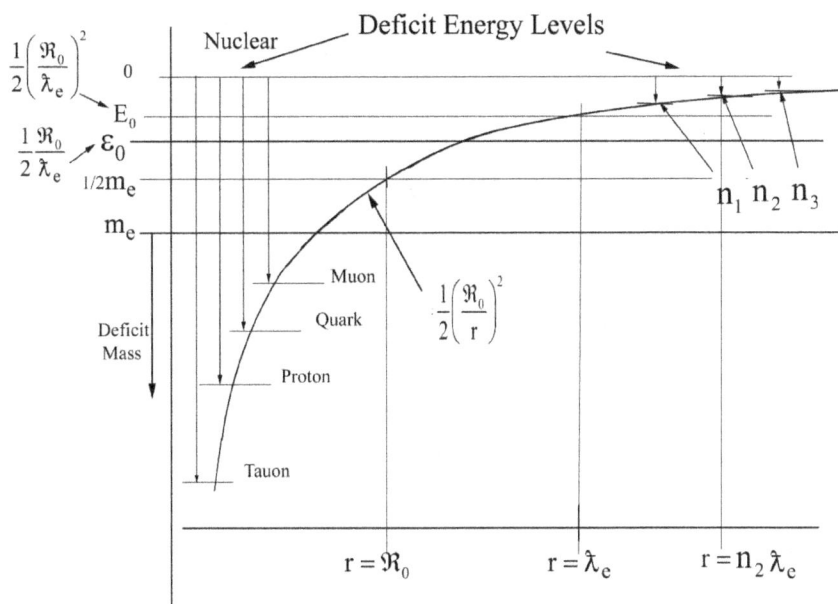

This graph illustrates the binding energy curve and the various state levels associated with the continuous binding energy function. When the binding energy exceeds ½ of the electron the energy is still positive. When the binding energy level exceeds the value of the electron the total energy of the particle becomes deficit energy. The states that exist within the nucleus have a deficit of energy.

Fig.3.

Nuclear Atomic Consilience

Atomic energy states are occupied by electrons having additional kinetic mass; the levels are identified by n and the particle's internal fine structure. Although they are identified by n numbers, each could be said to be a distinct particle and given a title. Energy levels in the nucleus are also occupied by particles, which are given names, with energy equal to the value of the mass of the particle and its internal fine structure.

Ground States and Nuclear-Atomic Boundaries

Particle mass and state energy have to be expressed with respect to their ground state. Energy levels in a system are referenced to the ground state and mass must be referenced to the mass ground state.

 There are two ground level boundary's associated with separating the positive and deficit states that are necessary to express the relation between states. Both boundaries occur at $r = \lambda_e$,

Potential Energy Ground State

The first is the Potential Boundary, defined by Δc Mechanics, which is the potential energy at the minimum radius between two free electrons E_0. It is defined as the lowest atomic level or the ground state and is only slightly lower than at the $n = 1$ Rydberg level atomic level. (See appendix II for distinction)

$$\frac{\Delta c}{c_0} = E_0 = \frac{1}{2}\left(\frac{\Re_0}{\lambda_e}\frac{\Re_0}{\lambda_e}\right) \quad \text{In fundamental units} \quad E_0 = \frac{Gm_e^4 c^3}{\hbar^3} \quad (19)$$

 This is the ground state of the potential energy, and all energy states for atomic as well as nuclear, are referenced to it.

Escape Energy Kinetic Boundary

The second boundary is the Positive-Deficit Kinetic Energy Boundary. Particles above this boundary have a positive total energy, and particles below this boundary have a deficit total energy. It is a boundary in a continuous function and can be found by noting that the binding potential is a scaled force between the particles, and the integral from its position at r to $r \rightarrow \infty$ is the escape energy, or the energy required to remove it from its bound position should the particle be stationary.

$$\frac{\varepsilon}{m_e c_0^2} = \int_r^\infty \frac{1}{2}\left(\frac{\Re_0}{r}\right)^2 = \frac{1}{2}\left(\frac{\Re_0}{r}\right) \qquad (20)$$

The boundary, ε_0, is at the same radius as that for the binding energy, $r = \lambda_e$, and the ground state kinetic Energy Boundary is:

$$\frac{\varepsilon_0}{\varepsilon_e} = \frac{1}{2}\left(\frac{\Re_0}{\lambda_e}\right) = \frac{1}{2}\frac{1}{136.718722887} \qquad (21)$$

Particles with Compton radii: "greater" than the electron, $\lambda > \lambda_e$ have greater kinetic mass than the deficit; whereas particles with Compton radii "less" than the electron have less kinetic mass than the deficit.

*Note that this paper's reference to m as energy implies $m \rightarrow mc_0^2$

The Binding of nuclear particles
See Nuclear Particle Structure in Δc Mechanics

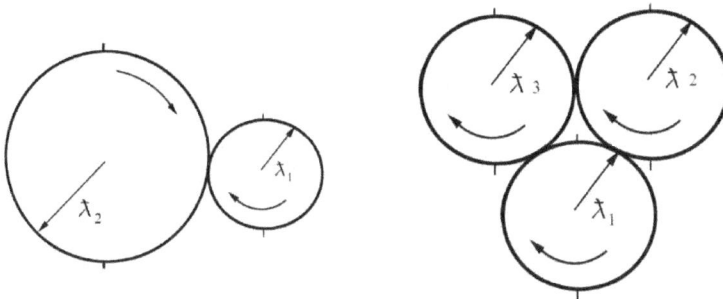

Conclusion

There are many items in this theory that need to be redone and, and there are many more that need to be done, but it is believed by the author that there is sufficient evidence presented here to show that it is worthy of serious investigation.

References:

1. DT Froedge, The Dirac Equation and the Two Photon Model of the Electron revised, April 2021, DOI:10.13140/ RG.2.2.19095.70564, https://www.researchgate.net/ publication/350922403

2. DT Froedge, The Connection between Electric Charge, Gravitation, and the Feynman Sum over All Histories View of Quantum Electrodynamics, April 2020 Conference: APS April. 18-21, 2020 Washington DC., https://absuploads.aps.org/ presentation.cfm?pid=18355 https://www.researchgate.net/ publication/341310206

3. DT Froedge, A Quantum Theory Conjecture on the Origin of Gravitational and Electric Particle Interaction, December 2019, DOI: 10.13140/RG.2.2.29097.54884 https://www.researchgate.net/publication/337826826

4. DT Froedge, The Electron as a Composition of Two Vacuum Polarization Confined Photons, April 2021, DOI: 10.13140/RG.2.2.18971.18722 , https://www.researchgate.net/ publication/350740864

5. DT Froedge, The Gravitational Constant to Eleven Significant Digits,
March 2020, DOI: 10.13140/RG.2.2.32159.38564 https://www.researchgate.net/publication/339943651

6. DT Froedge, The Fine Structure Constant from the Feynman Path Integrals, March 2021, DOI:10.13140/RG.2.2.12979.55846, https://www.researchgate.net/publication/350188862

8. DT Froedge, Vacuum Polarization, Gravitation, Charge, and the Speed of Light, Sept. 2021 DOI:10.13140/RG.2.2.15619.22569. https://www.researchgate.net/publication/354474157

9. DT Froedge, The Calculated value of the Fine Structure Constant from Fundamental Constants, September 2021 DOI:10.13140/RG.2.2.34349.41440

10. DT Froedge, Nuclear Particle Structure in Delta C Mechanics, September 2023, DOI: 10.13140/RG.2.2.25916.82562 https://www.researchgate.net/publication/374153059

11. Roger Blandford, Kip S. Thorne, Applications of Classical Physics, (in preparation, 2004), Chapter 26 http://pmaweb.caltech.edu/Courses/ph136/yr2012/1227.1.K.pdf

12. I. Shapiro; Gordon H. Pettengill; Michael E. Ash; Melvin L. Stone; et al. (1968). "Fourth Test of General Relativity: Preliminary Results". Physical Review Letters. 20 (22): 1265–1269. Bibcode:1968PhRvL..20.1265S. doi:10.1103/PhysRevLett.20.1265.

Appendix I

The Relation of the Fine Structure Constant

to the Electron Creation Radius And The

Anomalous Factor $g_A = (1 - g/2)$

The derived value of the fine structure constant in reference [6] from Eq.(31), and Eq.(32), of the paper is:

$$\alpha = \frac{\sqrt{2}\, \lambdabar_{PL}\, v_e}{\lambdabar_e\, g_A^2} = \frac{\mathcal{R}_0}{\lambdabar_e g_A^2} \qquad \lambdabar_{PL} = \sqrt{\frac{G\hbar}{c^3}} \qquad (22)$$

The value of the gravitational constant used here is from Quin BIPM 01-32-2001

$$G = 6.67559(27) \times 10\text{-}8 \text{ cgs}$$

In the paper on vacuum polarization [4], the orbiting radius and energy for two free photons in forming the electron was shown to

23

be $\mathfrak{R}_0 = \sqrt{2}\,\lambda_{PL}\,v_e$, which has been designated as Electron Creation Radius, and as such creates the mass of the electron as the lowest invariant mass in the universe.

This radius $\mathfrak{R}_0$ is a scalar invariant, and the variables are the Planck radius and the electron Compton frequency $v_e = c_0 / \lambda_e$:

$$\lambda_{PL} = \sqrt{G\hbar / c^3} = 1.616258265E - 33 \text{ cm} \tag{23}$$

$$v_e = 1.235589964E + 20 \text{ cycles / sec} \tag{24}$$

The value of $\mathfrak{R}_0$, is:

$$\mathfrak{R}_0 = 2.824230434E\text{-}13 \text{ cm}$$

The classical value of the radius of the electron is:

$$\alpha\lambda_e = 2.81793835793e - 13 \text{ cm} \tag{25}$$

The values of these two radii are very close, but the ratio is the square is the Anomalous gyromagnetic ratio.

$$\sqrt{\frac{\mathfrak{R}_0}{\alpha\lambda_e}} = 1.001159652181 = g_A \tag{26}$$

And for the Rydberg atomic levels this is:

$$\frac{m_K}{m_e} = \frac{1}{2}\left(\frac{\mathfrak{R}_0}{n\lambda_e}\right)^2 = \frac{1}{2}\left(\frac{\alpha}{n}\right)^2 g_A^4 \tag{27}$$

This difference is due to the value of Alpha has been defined inadvertently by incorporating the delay factor of the Feynman paths, g_A into its mass. The anomalous gyromagnetic ratio, thus increases the classical radius of the electron, and decreases the apparent mass whereas for $\mathfrak{R}_0$ is defined with just the Compton radius $\lambda_e = \hbar / m_e c_0$

$$
\begin{aligned}
\alpha &= 0.00729735253594 &= 1/137.03599971 \\
\frac{\mathfrak{R}_0}{\lambda_e} &= 0.007.3142871641 &= 1/136.7187229
\end{aligned}
\tag{28}
$$

Appendix III

Relativistic Mass for Two Opposite Moving or Orbiting Photons

The relativistic Doppler shift of opposite photons in the primed frame is:

$$v'_1 = v_1 \left[\frac{1 - \frac{v}{c}}{1 + \frac{v}{c}} \right] \qquad v'_2 = v_2 \left[\frac{1 + \frac{v}{c}}{1 - \frac{v}{c}} \right] \qquad (29)$$

Using the convention $m = p / c_0 = hv / c_0^2$, for photon mass:

$$m'_1 = m_1 \left[\frac{1 - \frac{v}{c}}{1 + \frac{v}{c}} \right] \qquad m'_2 = m_2 \left[\frac{1 + \frac{v}{c}}{1 - \frac{v}{c}} \right] \qquad (30)$$

Multiplying the two relations gives:

$$m'_1 m'_2 = m_1 m_2 = \text{constant} \qquad (31)$$

or:

$$m_1 m_2 = \frac{\left(m_1 + m_2 \right)^2 - \left(m_1 - m_2 \right)^2}{4} \qquad (32)$$

or:

$$\left[1 - \left(\frac{v}{c} \right)^2 \right] = \frac{4 m_1 m_2}{\left(m_1 + m_2 \right)^2} = \frac{m_0^2}{m^2} \qquad (33)$$

Thus the relativistic mass for two bound rotating photons is:

$$m \left[1 - \left(\frac{v}{c} \right)^2 \right]^{1/2} = m_0 \qquad (34)$$

The Lorentz Doppler shift change in the energy of the two photons in the electron transforms just as the relativistic mass.

Nuclear Particle Structure in Delta C Mechanics

D.T. Froedge

v 11 11 23

Abstract

The purpose of this paper is to illustrate the procedures of Δc Mechanics in calculating relative masses of the internal particles of the nucleus, including the mass relation of the proton to the primary leptons, the Z and W boson. The concept of Δc Mechanics which develops the binding mechanism for photons and particles in atomic systems extends into the nucleus where it is shown to be a continuing of the functions for mass particles greater than the electron. There is an extraordinary consilience between atomic and nuclear structure.

It has been found in earlier papers, that the electron is the ground state rest mass for atomic particles [8], thus it is not surprising that it is the ground state of particles in the nucleus.

The major difference is that the larger nuclear particles are occupants of states that have a greater binding energy than their total residual kinetic energy thus they are deficit energy particles.

Atomic particles that have kinetic mass greater than their binding energy are designated as positive energy particles, and nuclear particles that have less kinetic mass than their binding energy are designated as deficit particles.

Items noted

A. It is found that interacting particles in the nucleus have a highly accurate relationship between masses, and the relationship is maintained by a clockwork interaction of the particles similar to cogs on a wheel.

B. It is found that particles inside the nucleus have anomalous spin factors just as the electron, and the same summation delay of path integrals affect the internal bound mass of elementary particles just as the that of the electron..

C. A number of identity relationships between states have been discovered that allow evaluation of the internal binding mass of many nucleons, and allows some understanding of the relation between state levels. Since experimental mass levels are known to a high degree of precision, the bound values in particle interaction can be evaluated to a high degree of precision.

D. A nuclear particle represents a deficit of energy. When a particle in a system acquires a state value, the binding energy is radiated away by the photon emissions. This is the same for atomic particles, but for nuclear particles the binding energy radiated away is greater than the residual kinetic mass of the particle. That doesn't mean energy is negative, for it has been radiated out into the universe, but the residual particle represents a deficit, or an energy hole

E. It is found that for all mass particles the interaction of states requites the states to be referenced to the ground state of the system, that being the electron. The ratio of the mass of any particle to the electron whether free or bound is always a Lorentz invariant.

Energy Potential and Mass

In previous papers [1-9], the Feynman path integral probability for the relation between the change in c and the energies energy of two opposite electrons has been expressed as a change in the binding potential energy, as:

$$\frac{\Delta c}{c_0} = \frac{m_B}{m_e} = \frac{1}{2}\left(\frac{\mathfrak{R}_0}{r_1}\frac{\mathfrak{R}_0}{r_2}\right) = \frac{m_e}{m_e} - \frac{m_K}{m_e} \tag{1}$$

The last two terms are the difference in the mass of the free electron and the kinetic inertial mass of one of the bound electrons.

For an atomic system the binding energy solutions occur when values of $\alpha = \mathfrak{R}_0 / n\lambda_e$:

$$\frac{m_B}{m_e} = \frac{1}{2}\left(\frac{\mathfrak{R}_0}{r_1}\frac{\mathfrak{R}_0}{r_2}\right) \rightarrow \frac{1}{2}\left(\frac{\mathfrak{R}_0}{\lambda_e}\frac{\mathfrak{R}_0}{\lambda_e}\right) \rightarrow \frac{1}{2}\frac{\alpha^2}{n^2} \tag{2}$$

Thus the Lagrangian solution for an atomic particle is:

$$\frac{m_e}{m_e} - \frac{1}{2}\frac{\alpha^2}{n^2} = \frac{m_K}{m_e} \tag{3}$$

For a particle bound in a system the terms in parenthesis in Eq.(2), and the fine structure term α in, Eq.(3), represents the binding energy or the deficit energy when the photons are radiated away. The kinetic mass is then the residual mass of the electron. The energy of the absent photons still exists somewhere, but is not part of the inertial mass.

From earlier developments [4], in Eq.(1) $\mathfrak{R}_0$, is the Electron Creation Radius, $\mathfrak{R}_0 = \sqrt{2}\lambda_{PL}v_e$, It is the binding radius of two Planck photons when they form the electron, λ_{PL} is the Planck particle radius, and v_e is a unitless number equal to the Compton frequency of the electron. $v_e = \hbar / m_e c_0^2$

From the paper on vacuum polarization [4], the electron is the first Lorentz rest mass level in the universe and the ground state level for the mass of all other particles. The ratio of the electron, rest energy to the rest energy of other mass particles is the free state energy of that particle in reference to its ground state.

The binding energy is based on the probability density of Feynman photons rotating around the center of mass and there is no discontinuity if r is greater or less than $\mathfrak{R}_0$. If r becomes less than $\mathfrak{R}_0$, $r < \mathfrak{R}_0$, the kinetic mass becomes negative. The equation cannot be valid unless kinetic energy is injected into the system, or the binding energy is less than the kinetic mass energy.

$$\frac{1}{2}\left(\frac{\mathfrak{R}_0}{r_1}\frac{\mathfrak{R}_0}{r_2}\right) - \frac{m_K}{m_e} = \frac{m_e}{m_e} \tag{4}$$

Once energy is injected into the system, there can be solutions, but when the particle acquires a state, as in the case for atomic particles, the binding energy or part of it is radiated away. The particle is then left as a deficit particle, or in an energy hole.

$$\frac{1}{2}\left(\frac{\mathfrak{R}_0}{n\lambda_X}\frac{\mathfrak{R}_0}{n\lambda_X}\right) - \frac{m_K}{m_e} = \frac{m_e}{m_e} \tag{5}$$

The difference between the atomic particle and the nuclear particle is that for the atomic particle the deficit is less than the residual kinetic mass, whereas for the nuclear particle the deficit energy exceeds the kinetic mass energy of the residual particle.

In the nucleus, a particle that has acquired a state that is revolving around another particle retains some kinetic energy. That energy is not radiated away as state binding energy and the particle retains some mass in addition to the election. The Lagrangian expressed in Eq. (5), at integral values of $\hbar$ has state solutions that define the nuclear mass particles some of which are developed in the enclosed identities.

A graph showing the atomic and nuclear mass levels and the ground level values is presented in Fig. 1. This graph illustrates the continuation of the binding function from atomic into nuclear states.

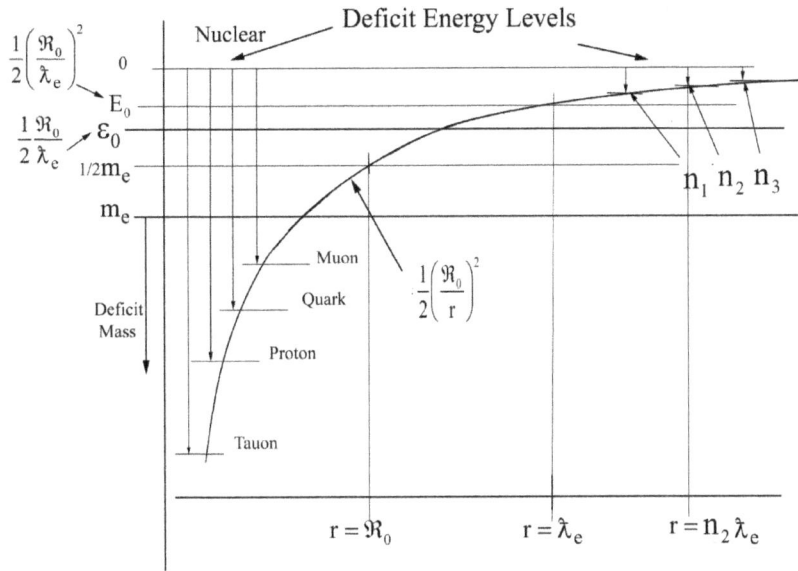

$$\frac{1}{2}\left(\frac{\mathfrak{R}_0}{\lambdabar_e}\right)^2$$

$$\frac{1}{2}\frac{\mathfrak{R}_0}{\lambdabar_e}$$

0

E_0

ε_0

$1/2m_e$

m_e

Deficit Mass

Muon

Quark

Proton

Tauon

Nuclear   Deficit Energy Levels

$$\frac{1}{2}\left(\frac{\mathfrak{R}_0}{r}\right)^2$$

$n_1\ n_2\ n_3$

$r = \mathfrak{R}_0$ $r = \lambdabar_e$ $r = n_2\lambdabar_e$

Fig.1.

This graph illustrates the binding energy curve and the various state level associated with the continuous binding energy function. When the binding energy exceeds ½ of the electron the energy is still positive. When the binding energy level exceeds the value of the electron the total energy of the particle becomes deficit energy, and there are can be states.

Nuclear Atomic Consilience

Atomic energy states are occupied by electrons having kinetic mass; the levels are identified by n and the particle's internal fine structure. Although they are identified by n numbers, each could be said to be a distinct particle and given a title. Energy levels in the nucleus are also occupied by particles, which are given names, with energy equal to the value of the mass of the particle and its internal fine structure.

Nuclear-Atomic Boundaries

There are two ground level boundary's associated with separating the positive and deficit states that can be defined and are useful for mechanics.

The first is the Potential Boundary, defined by Δc Mechanics, which is the potential energy at the minimum radius between two free electrons E_0. It is defined by the lowest atomic level and only

31

slightly lower than the n = 1 Rydberg level atomic level (See appendix x for distinction.). The boundary is at $r = \lambda_e$, thus:

$$\frac{\Delta c}{c_0} = E_0 = \frac{1}{2}\left(\frac{\mathcal{R}_0}{\lambda_e}\frac{\mathcal{R}_0}{\lambda_e}\right) \quad \text{In fundamental units} \quad E_0 = \frac{Gm_e^4 c^3}{\hbar^3} \quad (6)$$

This is the ground state of the potential energy, and all energy states for atomic as well as nuclear, are defined by reference to it.

The second boundary is the Deficit Energy Boundary or the escape energy of two particles, ε_E. This can be arrived at by noting that the binding potential is a force between the particles, and thus the integral from its position at r to $r \to \infty$ is the escape energy, or the energy required to remove from its bound position should the particle be stationary

$$\varepsilon = \int_r^\infty \frac{1}{2}\left(\frac{\mathcal{R}_0}{r}\right)^2 = \frac{1}{2}\left(\frac{\mathcal{R}_0}{r}\right) \quad (7)$$

The boundary is at the same minimum radius as that for the binding energy, λ_e, and the ground state kinetic Energy Boundary is:

$$\varepsilon_0 = \frac{1}{2}\left(\frac{\mathcal{R}_0}{\lambda_e}\right) \quad (8)$$

Particles with Compton radii: "greater" than the electron, $\lambda > \lambda_e$ have greater kinetic mass than the deficit; whereas particles with Compton radii "less" than the electron has less kinetic mass than the deficit.

*Note that this paper's reference to m as energy implies $m \to mc_0^2$

Relations between Nuclear States

For a particle solution with an x particle, $\lambda_x < \lambda_e$, the binding can be expressed as a product of E_0, and the binding energy of the particles.

$$\frac{1}{2}\left(\frac{\mathcal{R}_0}{n\lambda_x}\frac{\mathcal{R}_0}{n\lambda_x}\right) = \frac{1}{2}\frac{\lambda_e}{\lambda_K} \quad (9)$$

By factoring E_0:

$$\frac{1}{2}\left(\frac{\mathfrak{R}_0}{\lambda_e}\right)^2 \times \left(\frac{\lambda_e}{n\lambda_x}\right)^2 = \frac{1}{2}\frac{\lambda_e}{\lambda_K} \tag{10}$$

This represents an identity between the binding energy of the particles λ_x, and the kinetic mass of the resulting particle λ_K. Simplifying with the value of E_0, Eq.(6), this becomes:

$$E_0\left(\frac{\lambda_e}{n\lambda_x}\right)^2 = \frac{1}{2}\frac{m_K}{m_e} \quad \rightarrow \quad E_0 = \frac{\frac{1}{2}\left(\frac{n\lambda_x}{\lambda_e}\right)^2}{\frac{m_e}{m_K}} \tag{11}$$

In addition to being the particle ground state, E_0, is now be identified in Eq.(11), as the ratio of the binding to the kinetic energy of the two particles.

For illustration the relation between the free particle energy value and the bound energy value of the x particle as:

$$E_0 = \frac{\frac{1}{2}\left(\frac{n_2\lambda_x}{\lambda_e}\right)^2}{\frac{m_e}{m_K}} \rightarrow\rightarrow \frac{\frac{1}{\text{Deficit Binding Energy}}}{\frac{1}{\text{Invariant Kinetic Energy}}} \tag{12}$$

The Potential Boundary E_0 is the ratio of the binding energy to the kinetic energy of bound particles.

It is clear that the electron mass is the base energy state for particle systems. Without this as the defining reference none of the particle relations and none of the following identities are possible.

Larger masses

There are more massive particles, namely the W and Z bosons that have several Gev of kinetic mass, that can't be represented by Eq.(12).

When the mass energy exceeds the reciprocal of E_0 however there are identities that do show another level of mass particles that defining the W and Z bosons. The Identities appear with the second order power of the binding energy boundary

$$E_0 \rightarrow E_0^2 \qquad (13)$$

Thus the invariant kinetic mass is the binding of another level.

$$E_0 = \frac{\dfrac{1}{2}\left(\dfrac{n_2 \lambda_X}{\lambda_e}\right)^2}{E_0 \dfrac{\lambda_K}{\lambda_e}} \qquad \rightarrow \qquad E_0^2 = \frac{\dfrac{1}{2}\left(\dfrac{n_2 \lambda_X}{\lambda_e}\right)^2}{\dfrac{\lambda_K}{\lambda_e}} \qquad (14)$$

Another type of relation between the particles mass has been found that define the equivalence of states. That is the binding energy of two or more particles is equal to the binding of two or more other states. These are generally related to the permutations of the quarks, but also present in the relation between states of the pions π_0, $\pi_\pm$, shown in Id4.

$$\frac{1}{2}\left(\frac{n_2 \lambda_X}{\lambda_e}\right)^j = \frac{1}{2}\left(\frac{n_2 \lambda_Y}{\lambda_e}\right)^k \left(\frac{n_2 \lambda_Z}{\lambda_e}\right)^L \qquad (15)$$

The existences of these types of relations clearly indicate an integral clock like nature of the particle-particle interaction with the electrons mass as ground state.

Clockwork

An understanding of the mechanical structure of the electron gives insight into the physical structure of nuclear particles. The rotating photons in the electron create planes of probability flow density that when engaged with an opposite particle engage in planes and reduce the velocity of light in the interaction plane. Normally particles are thought of as existing in spherical structures, but particle interaction is the result of planar rotating electrons and their interaction is also planar.

The reduction of c occurs in the interaction zone of two self-bound planar photons rotating in the same direction but with the probability flow of their photons in opposite directions.

This interaction zone allows an exchange probability flow. Particles, or state solution to the interaction of two particles occur when the probability counter flow of the wavefunction of one particle satisfy counter flow continuity of the other. It creates the standing wave solutions to particle-particle interaction. This allows the Schrodinger equation to have perfect solutions for an atom even though it is not relativistically invariant. It could also allow two linear moving photons to form neutrinos.

The planes align and particles attract because the counter-flowing probability densities reduce the velocity of light (increase the index of refraction), thus inducing the particles attract. The planar rotation photon probability flow of the Feynman photons, as well as their centers attract, thus forming particles, or states aligned in planes.

For nuclear particles the same features apply, particles engage with other particles in planes, and thus define states that are other particles.

The above relates to two opposite particles, but states formed by three particles of equal or unequal mass would be the same. The proton with three quarks is illustrative.

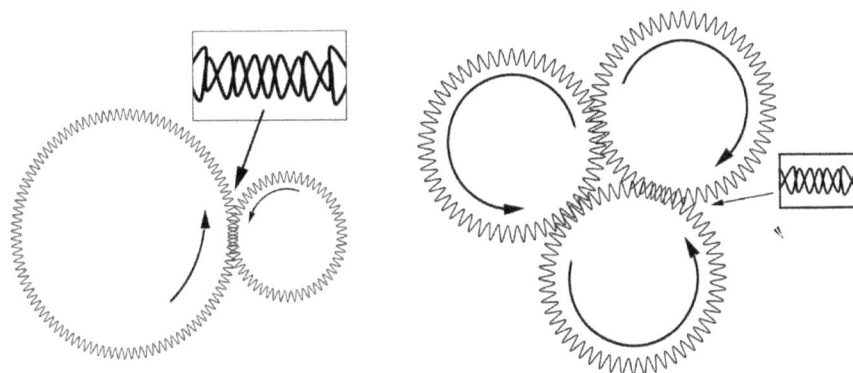

Although the particles are shown rotating clockwise, in the interaction zone the probability density is counter-flow creating a reduced value of c, or the index of refraction, and conjugate standing waves

Fig. 2

Note in Fig. 2, that when photons are rotation in opposite directions there is no counter flow of photons the volume of the space between the particles has no decrease in c, the reduction in c is on the exterior of the particles, causing them to be pulled apart. When particles are smaller than the electron the repulsion mechanism becomes very weak.

When particles engage, the frequency ratio between the particles is numerically an integer. That is the number of cycles of one has to be an integral number of cycles of the other. The engagement has clocklike precision. The engagement of the electron with a muon has to complete the number of cycles as well that of the tauon thus:

$$\frac{v_\mu}{v_e} = n_1 \quad \frac{v_\tau}{v_e} = n_2 \quad \rightarrow \quad \frac{v_\tau}{v_\mu} = \frac{n_2}{n_1} = n_3 \tag{16}$$

Given that, an identity relation between particles has to be precise to the number of cycles of each particle. Thus if the mass of two particles is known to parts per billion, the actual ratio has a precision much higher than that. This somewhat explains the precise relative mass of particles.

Showing the ratio of the frequencies of interacting particles are integers

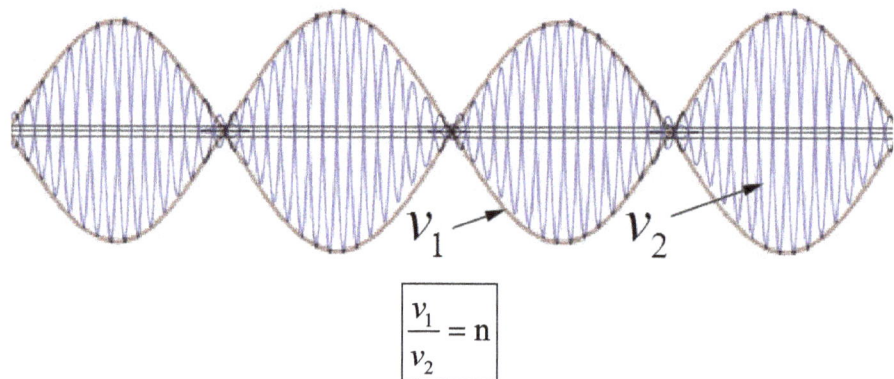

$$\boxed{\frac{v_1}{v_2} = n}$$

Fig. 3

Nuclear Particle Bound Mass vs Free Mass

There are differences in the experimental free mass of particles and the mass as indicated in the identities of bound mass enclosed it this paper the difference is unique for each particle and can be calculated to the accuracy of the experimental mass, which is accurate to parts per billion.

Initially this was thought to be a fine structure that is shown in atomic structure as a difference is spin and angular momentum. This turned out not to be the case, and is now thought to be related to the anomalous spin factor. This factor is not overtly presents in the atomic fine structure spectra.

$$g_A = (1 - g/2) = 1.001159652181 \qquad (1)$$

That is because the anomalous g factor and is independent of the differential atomic spectra values:

$$\frac{\Delta m_K}{m_e} = \frac{1}{2}\left(\frac{\Re_0}{n\lambda_e}\right)^2 = \frac{1}{2}\left(\frac{\alpha g_A^2}{n}\right)^2$$

$$f_e = \frac{m_e'}{m_e} = 1 - \left(\frac{\alpha g_A^2}{2}\right)^2 = \left(1 - \frac{1}{2}\frac{\Re_0}{\lambda_e}\right) = 1 - 0.00365714358240 \qquad (2)$$

Thus the Feynman QED factor for the change in mass reduces the mass of the electron as the result of being bound and is:

$$f_e = \frac{m_e'}{m_e} = 0.99634285642 \qquad (3)$$

The anomalous spin factor for the electron is the result of the Feynman path integral calculation of the change in the electrons effective mass as the result of orbital motion. As the election moves in a circle the delay of the probability density in effect slows down the orbital motion decreases the effective mass and, increases the radius. This change is not present in the free particle mass but it is present when the particle is in a bound state. Since it defines the ground state it is not part of the differential or spectral values, and does not show up in spectral values.

Nuclear particles bound in the nucleus have exactly the same type of factor

For nuclear particles there is a segue into very accurate values of its magnitude. For particles bound together as clockwork systems, the Identities must be exact for the mass ratio of each of the bound particle. By using the calculated residuals in the identities, and the matrix of the number of particles, a unique exact value for each particle mass can be found.

$$f_X = \frac{m_X'}{m_X} \qquad (4)$$

The ratio of the free particle mass, m_X to the bound mass found in the identity relations fives the f factor associated with that particle, f_X gives a mass in the identities that exactly satisfies the clockwork relations between the nuclear particles. A single value of f for a particle mass is the same in all the identities shown.

Procedure

Since most of the experimental mass values of the elementary particles are known to accuracies in parts per billion, the mass ratios for bound vs free particles can be found to the same accuracy of parts per billion.

The included identities, (17) have a variety of mass relations that that give the state values, state value ratios, and mass ratios of bound particles.

The mass of the particles as determined from the bound particle identities are not the same as the experimentally determined free mass value, the mass relations are slightly different from the free particle value. Because of the clockwork interaction mechanism of nuclear particles noted earlier these relations are highly accurate and probably more accurate than rest mass measurements. Inserting the rest mass values into these relations returns residuals that can be used to determine the bound mass value of the particles. By treating the identities as a matrix of equations with unique valued particles each particle's bound mass can be calculated to a high accuracy

The synthesized values are included in Table I.

Table I

Particle	Codata Mass (gm)	Free Electron Mass Ratio	Feynman f Factor f_x	Bound Mass Electron Ratio
Electron	9.10938370150E-28	1.000000000	0. 9963428564	0. 996342856
Muon	1.88353162700E-25	206.768282984	0.99879668589	206.5194757920
Quark n=8	2.09077740461E-25	229.519084180	1.00240275088	230.0705613606
Proton	1.67262192369E-24	1836.152673440	1.00180152226	1839.4605433540
Tauon	3.16754000000E-24	3477.188250779	1.00360629040	3489.7280013871
Neutron	1.67492749804E-24	1838.683661732	1.00060117395	1839.7890304522
Neutral Pion	2.40618001661E-25	264.143008513	0.99310725370	262.3223377684
Charged Pion	2.48806819638E-25	273.132439900	1.00111669593	273.4374457843

Surprisingly, the mass ratio of the free particle to the bound value for the four of the particles, the muon, tauon, proton, and quark are mathematically related, and the common factor is

There is a common sub-factor in the f factor for the primary particles. Those particles are the muon, tauon quark and the proton, and the value of the f factors is powers of that sub-factor that sub-factor are:

$$sf = 0.001199934442093 \qquad (5)$$

For the different bound particles in the nucleus it is remarkable that they are be defined by the same number. Because of the relation of the free particle mass with the bound values is expected that this number will be found the methods of path integrals

The relations between those particles mass is:

$$f_\tau^{3} = f_P^{2} = f_Q^{3/2} = f_\mu^{-1} \qquad (6)$$

Combining those relations with the sub-factor above gives the four f functions shown are in Table II.

Table II

f_X *Factor*

Particle	Bound to Free Mass Ratio	Formula		Value
Tauon	$f_\tau = \dfrac{m'_\tau}{m_\tau} =$	$e^{3 \times \left(0.001199934442093\right)}$	$\rightarrow$	$f_\tau = 1.00360629040$
Quark n = 8	$f_Q = \dfrac{m'_Q}{m_Q} =$	$e^{2 \times \left(0.001199934442093\right)}$	$\rightarrow$	$f_Q = 1.0024027508760$
Proton	$f_P = \dfrac{m'_P}{m_e} =$	$e^{3/2 \times \left(0.001199934442093\right)}$	$\rightarrow$	$f_P = 1.00180152238149$
Muon	$f_\mu = \dfrac{m'_\mu}{m_\mu} =$	$e^{-\left(0.001199934442093\right)}$	$\rightarrow$	$f_\mu = 0.99879668589304$

The f factor values expressed in the right column are in agreement within the group of identities to at least 1 part in 4 billion, except the muon is smaller than the formula by about one part in 250,000

Figure 3 shows the apparent shift of the particle position with respect to the boundary mass $1+\varepsilon_0 = 1 + \dfrac{1}{2}\dfrac{\Re_0}{\lambda_e} = 0.003657143582405$, as the result of the internal structure. Graphically the values can be expressed as if they are shifts from the Mass Boundary

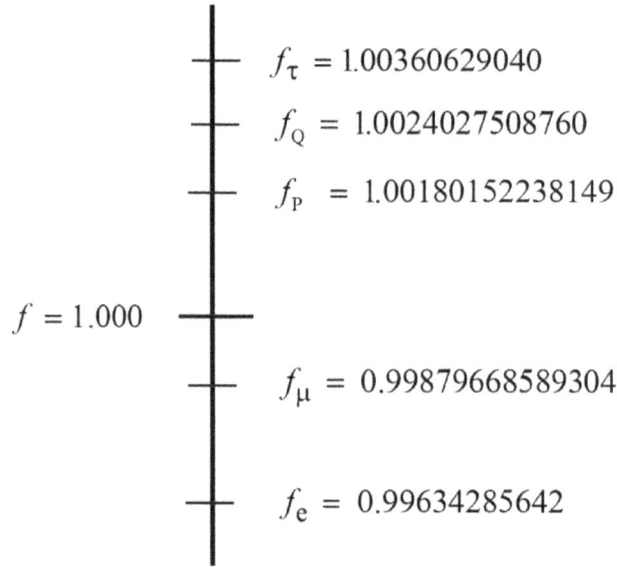

$$f_\tau = 1.00360629040$$

$$f_Q = 1.0024027508760$$

$$f_P = 1.00180152238149$$

$$f = 1.000$$

$$f_\mu = 0.99879668589304$$

$$f_e = 0.99634285642$$

Fig. 3.

Identities

There has been found a number of identities relating the particle masses of various particles with a high degree of accuracy. Given more effort, there are undoubtedly many more that could be found. The 17 identities illustrated in this paper are enough to develop some useful insight, and calculate the bound mass values for several particles. They also give insight into the makeup of particle state levels and can assist in developing a comprehensive Lagrangian for nuclear particles.

It is presumed that all particles are constructed of three primary leptons, the electron, the muon, the tauon, and perhaps the quark. The reference mass for the particles defined here is the electron, the first invariant rest mass level in the universe [4].

The primary variables in the following Identities originate with the fundamental constants:

$$\mathfrak{R}_0 = \sqrt{2}\, \lambdabar_{PL} \nu_e \qquad \lambdabar_{PL} = \sqrt{\frac{G\hbar}{c^3}} \qquad \alpha = \frac{\sqrt{2}\, \lambdabar_{PL} \nu_e}{\lambdabar_e\, g_A^2} = \frac{\mathfrak{R}_0}{\lambdabar_e g_A^2} \qquad (7)$$

These are the Electron creation radius, $\mathfrak{R}_0$, the Planck particle radius, λ_{PL} and the fine structure constant, α.

Unless otherwise noted all masses are expressed in ratios of the free electron mass. The first table is a list of constants and particle masses based on the Codata recommended experimental values; this data allows insight into some of the operating rules for internally bound particles with other bound particles. There are still some unresolved issues as to particle configurations. The calculations are made in excel with about 15 significant digits, and no tracking of the relative uncertainty in the calculations has been made.

The variables λ_e, λ_τ, λ_P, and λ_Q, are the Compton radii $\lambda = \hbar \, / \, mc_0$, of the electron, tauon, proton, and quark.

The Identities show:

Particle mass states as a ratio of the electron mass have integral values of energy relations that define a multitude of particles as solutions.

Mass values are defined for the primary leptons, the proton and the quarks, and the relative values for the W and Z boson.

There are four general types of identities in the following list :

Relative:

These are identities that give ratios of particle states in the nucleus

State:

Identities that give the bound value of nuclear particles reference to the ground state, E_0.

Second stage:

Identities that show the next step to heaver elementary of elementary particles is defined, E_0^2, W and Z Boson.

Heavy particles:

Identities show the W and Z boson relation to the proton, tauon, and muon, and the Top quark,

Identity Id2, the Tauon Proton Relation

The identity relation between the tauon and the proton has been found to be similar to the tauon, proton relation:

$$\frac{\left(\dfrac{2^7 \lambda'_P}{\lambda_e}\right)^2}{\left(\dfrac{\lambda'_\mu}{\lambda_e}\right)} = 1 \tag{8}$$

Identity Id1, the Tauon Quark Relation

The concept of Δc Mechanics as related to massive particles can be best illustrated by the identity relation between the tauon and the proton.

$$\frac{1}{2}\left(\frac{\lambda_\tau}{\lambda_e}\frac{\lambda_\tau}{\lambda_e}\right) = \frac{1}{2}\left(\frac{2^3 \lambda_P}{\lambda_e}\right)\left(\frac{2^3 \lambda_P}{\lambda_e}\right)\left(\frac{2^3 \lambda_P}{\lambda_e}\right) \tag{9}$$

The mass of the proton is known to a greater precision than the tauon, and presuming Eq.(9), is valid, the tauon mass, can be calculated to the same precision as the proton with a much reduced uncertainty.

Codata Proton mass in electrons with uncertainty bars   1836.1526734(56) electrons
Codata Tauon mass in electrons with uncertainty   3477.22(23) electrons
Tauon mass result from Eq.(9), with proton uncertainty   3477.188250(16) electrons

Tauon Quark Proton Connection

For three quarks with $n = 2^3 = 8$ the counter rotation forms a standing waves at the intersection satisfying the flow continuity relations for particle interaction Fig. X.

In Eq.(9), the value of each of the right side terms is 1/8 the mass of the proton and is 1/229.519. The three quark terms shown with 2^3 can be expressed as

$$\left(\frac{\lambda_\tau}{\lambda_e}\right)^2 = \left(2^{n1} A \times 2^{n2} B \times 2^{n3}\right) = Q_1 Q_2 Q_3 \tag{10}$$

Permutations of the total product $8^3 = 512$ between three particles can be identified with the multiplicity of the quark masses.

The powers of 2^n distributed among three particles define 12 permutations of three particles. It is asserted that these permutations constitute the 12 quark leptons; the 6 quarks and their 6 anti-quarks. There are three nondegenerate states, and nine states with multiple degeneracies. . (See Appendix I)

For this relation to exist, an integral number of cycles of the tauon must be exactly equal to an integral number of cycles of the electron, and each quark likewise.

Elementary Particle State values

From the identities, some of the nuclear state value can be found for atomic particles, particles which ae in effect state values in the nucleus. They are referenced to the ground state boundary identified in Eq.(6), E_0

$$\left(\frac{1}{2^8}\frac{\lambda'_\mu}{\lambda_e}\right) = \left(\frac{E_0}{\sqrt{2}}\right)$$

$$\left(\frac{\lambda'_\tau}{\lambda_e}\right)^{\frac{4}{3}} = \left(\frac{E_0}{\sqrt{2}}\right)\frac{1}{f_\mu}$$

$$2.6749398365E-05 \qquad\qquad 2.6781625072E-05 \qquad (11)$$

$$\left(\frac{2^3\lambda'_P}{\lambda_e}\right)^2 = \left(\frac{E_0}{\sqrt{2}}\right)$$

$$\left(\frac{\lambda'_Q}{\lambda_e}\right)^2 = \left(\frac{E_0}{\sqrt{2}}\right)\frac{1}{f_\mu}$$

Note that the primes on the radius indicate that the mass in the nucleus includes the offset of the level due to the internal f factor of the particle. The mid numbers in Eq.(11), are the value of each factor. These relations are accurate to parts per billion.

The states are highly accurate as shown in the identities. The states on the left Eq.(11), appear to include the additional f factor offset of the muon.

Particle Data Table

Particle	m_x Particle Mass (gm)	λ_e / λ_x Mass Electron Ratio	λ_x Compton Radius (cm)	f_x Feynman f Factor Exact	$\dfrac{m_x}{m_e} \times f_x$ Mass Including Fine Structure
Electron	9.10938370150E-28	1.000000000	3.86159267944E-11	0.99634285641760	0.9991011272
Muon	1.88353162700E-25	206.768282984	1.86759430591E-13	0.998796685893035	206.5194757920
quark n=8	2.09077740461E-25	229.519084180	1.68247128261E-13	1.002402750876000	230.0705613606
Proton	1.67262192369E-24	1836.152673440	2.10308910326E-14	1.001801522381490	1839.4605435770
Tauon	3.16754000000E-24	3477.188251050	1.11055036504E-14	1.003606290324000	3489.7280013948
Neutron	1.67492749804E-24	1838.683661732	2.10019415510E-14	na	na
Neutal Pion	2.40618001661E-25	264.143008513	1.46193257250E-13	0.993107253795974	262.3223377938
Charged Pion	2.48806819638E-25	273.132439900	1.41381693103E-13	1.001116695968690	273.4374457949
CTR Radius $\mathfrak{R}_0$			2.82447977710E-13		
Planck Particle Radius			1.61640095996E-33		

Fine Structure Constant α	0.007297353	$1/\alpha$	137.0359997	
Planck Constant $\hbar$	1.054571817600E-27 (cgs)			
Velocity of light	29979245800 (cm/sec)			
Electron Frequency	1.23558996386E+20 (hz)			
Electron mass	510998.95 (ev)			

$$E_0 = \frac{1}{2}\left(\frac{\mathfrak{R}_0}{\lambda_e}\right)^2 \qquad \frac{1}{2}\frac{\mathfrak{R}_0}{\lambda_e}$$

2.67493983647E-05	0.0036571436
37384.0184	0.0146285743

Particle-Particle Identities
State-State Relations

Id	Relation	Factors / Reciprocals		Ratio of Main Factors

Id 1

$$\frac{1}{2}\left(\frac{\lambda'_\tau}{\lambda_e}\right)^2 = \frac{1}{2}\left(\frac{\lambda'_Q}{\lambda_e}\right)^3$$

Particle in Parenthesis = quark

Factors	Reciprocals	Ratio of Main Factors
4.1056965515E-08	24356403.05	1.000000000000000
4.1056965515E-08	24356403.05	

Id 2

$$\frac{1}{2}\left(\frac{2^7 \lambda'_P}{\lambda_e}\right)^2 = \frac{1}{2}\left(\frac{\lambda'_\mu}{\lambda_e}\right)$$

Particle in Parenthesis = not quark

Factors	Reciprocals	Ratio of Main Factors
4.8421583299E-03	206.5194757920	1.000000000000000
4.8421583299E-03	206.5194757920	

Id 4

$$\frac{1}{2}\left(\frac{\lambda'_\tau}{\lambda_e}\right)^2 = \frac{1}{2}\left(\frac{\lambda'_\mu}{\lambda_e}\right)^2\left(\frac{2\lambda'_{\pi+}}{\lambda_e} - \frac{\lambda'_{Pi0}}{\lambda_e}\right)$$

Factors	Reciprocals	Ratio of Main Factors
4.105696551549E-08	24356403.04744	1.000000000155790
4.105696552188E-08	24356403.04364	

Id 5

$$\frac{\lambda'_{\pi+}}{\lambda_e} = \frac{1}{2}\left(\frac{\mathfrak{R}_0}{\lambda_e}\right) = \varepsilon_0$$

Factors	Reciprocals	Ratio of Main Factors
3.6571435821E-03		1.000000000077890
3.6571435824E-03		

The first identify Id1 is the binding energy of two tauons shown equal to the state value of three bound quarks, discussed earlier. When three particles are bound, the binding energy is proportional to the product of the mass ratios.

Id2 is between bound protons being related to the bound state of the muon. Calculations show that the proton in this relation is not a quark but bound protons. The free mass of a primary quark is 1/8 of the free mass of the proton, but the f factor for the quark is not 1/8 of the f factor of the proton, but has its own f factor.

Id3 shows the state value of two bound tauons being equal to, the difference between a charged and an uncharged pion bound with a muon.

Id4 shows the bound value of the charged pion is equal to the ground state of particle kinetic mass, mass ε_0.

$\boxed{E_0}$ Particle Mass Definition States   Ratio of Main Factors

Id 6	$\left(\dfrac{2^3\,\lambda'_P}{\lambda_e}\right)^2 = \dfrac{E_0}{\sqrt{2}}$	2.6749398365E-05 2.6749398365E-05		1.000000000000000
Id 7	$\dfrac{1}{2^8}\dfrac{\lambda'_\mu}{\lambda_e} = \dfrac{E_0}{\sqrt{2}}$	2.6749398365E-05 2.6749398365E-05	37384.01838 37384.01838	1.000000000000010

$\boxed{E_0^2}$

Id 8	$\dfrac{\dfrac{1}{2}\left(\dfrac{\lambda'_P}{\lambda_e}\right)^2}{\dfrac{\lambda_e}{\lambda'_\mu}} = E_0^2$	1.47770945127E-07 7.15530312871E-10	206.519475792 206.519475792	1.000000000000010
Id9	$\dfrac{\dfrac{1}{2}\left(\dfrac{\lambda'_\mu}{\lambda_e}\right)^2}{2^{14}} = E_0^2$	7.1553031287E-10 7.1553031287E-10		1.000000000000010

These identities 6-9 show energy levels of the nuclear particle mass and the ground state E_0, the state that sets the mass of nuclear particles. The second power of the ground state E_0^2 is the second level of particle mass, for the heavier mass particles Id12, and Id16.

Proton State Definitiion

Id10
$$\frac{\frac{1}{2}\left(\frac{\lambda'_\mu}{\lambda_e}\right)^2}{\left(\frac{2^9 \lambda'_\tau}{\lambda_e}\right)^2} = \frac{\lambda_P}{\lambda_e}$$

1.1723248646E-05 1836.152672868

2.1525674336E-02 Exp. 1836.15267344 1.000000000311550

Id11
$$\frac{\frac{1}{2}\left(\frac{2^2 \lambda'_\tau}{\lambda'_{\pi+}}\right)^2}{E_0} = \frac{\lambda_e}{\lambda_P}$$

4.9115979313E-02 1836.15267315

2.6749398365E-05 Exp. 1836.15267344 1.000000000155780

Id 12
$$\frac{\left(\frac{2^2 \lambda'_\tau}{\lambda_e}\right)^2}{E_0^2} = \frac{\lambda_e}{\lambda_P}$$

1.313822896E-06 1836.15267287

7.155303129E-10 Exp. 1836.15267344 1.000000000311560

These identities 10-12 show the state values that define the state mass of the proton. The proton is the free mass not the bound value, and thus not a participant in the particles that define the state.

Z Boson State Dfinition

Z Boson relation with Proton and ground state

Id 13

$$\frac{\frac{1}{2}\left(\frac{\lambda_z}{\lambda_e}\right)^2}{\frac{1}{2}\left(\frac{2\lambda_p}{\lambda_e}\right)^2} = E_0$$

$\boxed{\lambda_z/\lambda_e}$ 5.6334989226E-06   1/ Electrons

177509.6 Electrons

510998.95 Volts/ Electron

$\boxed{m_z}$ 90707037849.86   90.70 Gev

1.0006

CERN ATLAS detector   90.76 Gev

Reference [14]

Decay Channel	Entries	Center	Error	Center deviance from Z_m^0(%)	Z-score	Width	Error
Muon	600071	90.76	(6.03E-03)	-0.47	70.70	5.71	(2.38E-02)
Electron	375216	90.04	(8.90E-03)	-1.26	128.96	6.57	(3.68E-02)

W Boson State Dfinition

W Boson relation with Proton and ground state

Id 14

$$\frac{\frac{1}{2}\left(\frac{2\lambda_p}{\lambda_e}\right)^2}{\frac{\lambda_w}{\lambda_\tau}} = E_0$$

6.3548721889E-06   = 1/ Electrons

$\boxed{m_W}$ 157359.58 Electrons

510998.95 Volts / Electron

$\boxed{m_W}$ 80410555053.28   80.41 Gev

1.00029

Reference [15]   Fermilab (CDF) Collabora 80.433Gev +/-9Mev/cc

W Boson relation with Tauon and Muom

Id 15

$$\frac{\lambda_w}{\lambda_e} = E_0 \frac{4\lambda_\tau}{\lambda_\mu'}$$

$\boxed{m_W}$ 6.354872189E-06   1/n Electrons

157359.58 Electrons

510998.95 Volts / Electron

$\boxed{m_W}$ 80410578657.21   80.41 Gev mass

1.00029

Reference [15]   Fermilab (CDF) Collabora 80.433Gev +/-9Mev/cc

Identity 13 notes that the ratio of the binding potential energy of two Z Bosons to the binding potential of two protons is E_0, the Potential Boundary between positive and deficit mass particles or the ground state of the potential.

Identity Id14 shows the Ratio of the binding energy of two protons to be the ¼ the ratio of the kinetic mass of the tauon to the W boson.

W and Z boson state Relation

Id 16

$$\frac{\frac{1}{2}\left(\frac{\lambda_Z}{\lambda_e}\right)^2}{\frac{\lambda_W}{\lambda'_\tau}} = E_0^2$$

7.1553031282E-10   Mass Comparison
7.1553031287E-10   calculated from   1.000000000077840
                   From Id14 & Id15

Top Quark Mass related to ground state

Id 17

$$\left(\frac{3^2\lambda_{TOP}}{\lambda_e}\right) = E_0$$

m_{TQ} .97215537385E-06 =1 / Electrons
336456.16538029 Electrons   1.00093
510998.95000000 Volts / Electron
m_{TQ} 171928747230.36   171.93 Gev

Refference [16]   CER CMS collaboration 171.77+/-0.38 Gev/cc

Identity 15 is a second relation between the W boson. The muon and the Tauon

Identify ID16 is a combination of Id 13 and Id 14 using the mass calculated for each

Top Quark Mass related to ground state

Id 17

$$\left(\frac{3^2\lambda_{TOP}}{\lambda_e}\right) = E_0$$

m_{TQ} .97215537385E-06 =1 / Electrons
336456.16538029 Electrons   1.00093
510998.95000000 Volts / Electron
m_{TQ} 171928747230.36   171.93 Gev

Refference [16]   CER CMS collaboration 171.77+/-0.38 Gev/cc

Id 17 is a relation showing the value of the mass of the top Quark in the n = 9 position to be equal to the ground state of atomic particles.

Conclusion

The model nucleon structure found by way of Δc Mechanics is a new approach to fundamental particle physics. It relies at finding the fundamental mechanisms that bind particles together .The first binding is of the Planck photons, creating the electron and binding electron, to form the larger particles. The electron, is the first rest mass energy level of the universe, and is the fundamental ground state mass.

The state values and relations in the identities found and listed here are somewhat random. But give insight into the structure and consilience between nuclear and atomic particles. The Identities illustrate the interactions of particles to be somewhat like clockwork or cogs. The clockwork interaction of the particle masses makes the precision of the defined masses explainable. Without this precision, particle masses measured in different parts of the world would, not agree to the known precision.

The deficit nature of mass particles may have a lot to contribute to the dynamics of the universe and the issues involving dark matter and dark energy. The mass, energy ratio in the universe may not be as straightforward as previously presumed.

Future

An important future consideration is developing the wave function for the muon and the tauon and perhaps the quark which also appears to be a two photon particle.

The electron is a creation of the vacuum polarization and the interaction of two Planck particles, and it is clear that the relations between all mass particles are defined relative to that mass. The method of combining of the photons to form the election in [1], was shown to be the magnitude of the four differentials, or four-momentum, of the two electron photons. This is the starting point for the wavefunctions of the two other primary leptons.

The electron is well defined in exponentials of $SU(2)$ [1], and thus has solutions that yield precise and useful solutions regarding atomic energy states. It's certain that the standard model success will require the muon and tauon wavefunctions to necessarily be defined in differentiable exponentials of $SU(2)SO(2)$

With those wavefunctions using the procedures developed in [1] and the nuclear Lagrangian, Eq.(4), and the analytical value of the fine structure factors can be developed.

Most of the understanding of the nuclear mechanics presented in this paper has been by discovery and exploration. But a more analytical approach to develop the details and connections to the mathematics of the standard model needs to be undertaken. The connection is quite apparent but there is a long way to go in incorporating the fundamental concepts.

Authors Note

Before now the most compelling attribute of the Δc mechanical theory was the calculation of the Fine Structure Constant in terms of fundamental constants, [4&8]. This paper is a logical extension of that theory, and the agreement with experimental values is particularly compelling. Accept for cleaning up some previous work, his will be the author's last paper.

DT Froedge 8/6/2023

References

1. DT Froedge, The Dirac Equation and the Two Photon Model of the Electron revised, April 2021, DOI:10.13140/ RG.2.2.19095.70564, https://www.researchgate.net/ publication/350922403

2. DT Froedge, The Connection between Electric Charge, Gravitation, and the Feynman Sum over All Histories View of Quantum Electrodynamics, April 2020 Conference: APS April. 18-21, 2020 Washington DC., https://absuploads.aps.org/ presentation.cfm?pid=18355 https://www.researchgate.net/ publication/341310206

3. DT Froedge, A Quantum Theory Conjecture on the Origin of Gravitational and Electric Particle Interaction, December 2019, DOI: 10.13140/RG.2.2.29097.54884
https://www.researchgate.net/publication/337826826

4. DT Froedge, The Electron as a Composition of Two Vacuum Polarization Confined Photons, April 2021, DOI: 10.13140/RG.2.2.18971.18722 , https://www.researchgate.net/publication/350740864

5. DT Froedge, The Gravitational Constant to Eleven Significant Digits,
March 2020, DOI: 10.13140/RG.2.2.32159.38564
https://www.researchgate.net/publication/339943651

6. DT Froedge, The Fine Structure Constant from the Feynman Path Integrals, March 2021, DOI:10.13140/RG.2.2.12979.55846,
https://www.researchgate.net/publication/350188862

8. DT Froedge, Vacuum Polarization, Gravitation, Charge, and the Speed of Light, Sept. 2021 DOI:10.13140/RG.2.2.15619.22569.
https://www.researchgate.net/publication/354474157

9. DT Froedge, The Calculated value of the Fine Structure Constant from Fundamental Constants, September 2021 DOI:10.13140/RG.2.2.34349.41440

8. M. Khodaverdian, Accuracy and Precision of the Z Boson Mass Measurement with the ATLAS Detector, https://indico.cern.ch

9 CODATA values of the fundamental physical constants, https://www.nist.gov/programs-projects/codata-values-fundamental-physical-constants

7. DT Froedge, Electron Mass and State Energy Levels Resulting from Photon-Photon Interaction, Conference APS Apr. 2022. Conference: APS April 2022, New York, https://www.researchgate.net/publication/359912763

11. P. Halpern, The Quantum Labyrinth: How Richard Feynman and
John Wheeler Revolutionized Time and Reality, Basic Books, New York, 2017

12. R. Feynman - Nobel Lecture: The Development of the Space-Time View of Quantum Electrodynamics, 1965 https://www.nobelprize.org/prizes/physics/1965/feynman/lecture/

13. K. Thorne, John Archibald Wheeler1911—2008 A Biographical Memoir, https://arxiv.org/ftp/arxiv/papers/1901/1901.06623.pdf

14 M. Khodaverdian, Accuracy and Precision of the Z Boson Mass Measurement with the ATLAS Detector, 27 05, 2019,

https://indico.cern.ch/event/813935/contributions/3557802/attachments/1919010/3174010/Gymnasieprojekt_Mariam_Khodaverdian_2019.pdf

15. CDF Collaboration, High-precision measurement of the W boson mass with the CDF II detector. Apr 2022.DOI: 10.1126/science.abk1781, https://www.science.org/doi/10.1126/science.abk1781

16. CMS collaboration, A profile likelihood approach to measure the top quark mass in the lepton + jets channel, April, 2022, https://home.cern/news/news/physics/cms-measures-mass-top-quark-unparalleled-accuracy

Appendix I

Permutations of the quarks

Permutations and Degeneracies of $2^A \times 2^B \times 2^C$

	A	B	C	#
z particle	9	0	0	1
	0	0	9	
meson	8	1	0	2
	1	8	0	
	8	0	1	
	0	8	1	
	1	0	8	
	0	1	8	
meson	7	2	0	3
	2	7	0	
	7	0	2	
	0	7	2	
	2	0	7	
	0	2	7	
meson	6	3	0	4
	3	6	0	
	6	0	3	
	0	6	3	
	3	0	6	
	0	3	6	
meson	5	4	0	5
	4	5	0	
	5	0	4	
	0	5	4	
	4	0	5	
	0	4	5	
3quark	7	1	1	6
	1	7	1	
	1	1	7	

	A	B	C	#
3quark	6	2	1	7
	2	6	1	
	6	1	2	
	1	6	2	
	2	1	6	
	1	2	6	
3quark	5	3	1	8
	3	5	1	
	5	1	3	
	1	5	3	
	3	1	5	
	1	3	5	
3quark	4	4	1	9
	4	1	4	
	1	4	4	
3quark	5	2	2	10
	2	5	2	
	2	2	5	
3quark	4	3	2	11
	3	4	2	
	4	2	3	
	2	4	3	
	3	2	4	
	2	3	4	
non degeren	3	3	3	12

Appendix II

The Relation of the Fine Structure Constant to the Electron Creation Radius And The Anomalous Factor $g_A = (1 - g/2)$

In the paper on vacuum polarization [4], the orbiting radius and energy for two free photons in forming the electron was shown to be $\Re_0 = \sqrt{2}\, \lambda_{PL} \nu_e$, which has been designated as Electron Creation Radius, and as such creates the mass of the electron as the lowest invariant mass in the universe.

This radius $\mathfrak{R}_0$ is a scalar invariant, and the variables are the Planck radius and the electron Compton frequency $\nu_e = c_0 / \lambda_e$:

$$\lambda_{PL} = \sqrt{G\hbar / c^3} = 1.616258265E - 33 \text{ cm} \tag{12}$$

$$\nu_e = 1.235589964E + 20 \text{ cycles / sec} \tag{13}$$

The value of $\mathfrak{R}_0$, is:

$$\mathfrak{R}_0 = 2.824230434E\text{-}13 \text{ cm}$$

The classical value of the radius of the electron is:

$$\alpha \lambda_e = 2.81793835793e - 13 \text{ cm} \tag{14}$$

The values of these two radii are very close, but the ratio is the square is the Anomalous gyromagnetic ratio

$$\sqrt{\frac{\mathfrak{R}_0}{\alpha \lambda_e}} = 1.001159652181 = g_A \tag{15}$$

And for the Rydberg atomic levels of hydrogen this is:

$$\frac{\Delta m_K}{m_e} = \frac{1}{2}\left(\frac{\mathfrak{R}_0}{n\lambda_e}\right)^2 = \frac{1}{2}\left(\frac{\alpha}{n}\right)^2 g_A^4 \tag{16}$$

This difference is due to the value of Alpha has been defined inadvertently by incorporating the delay factor of the Feynman paths, g_A into the electron mass. The anomalous gyromagnetic ratio, increases the classical radius of the electron, and decreases the apparent mass whereas $\mathfrak{R}_0$ is defined with just the Compton radius $\lambda_e = \hbar / m_e c_0$

$$\alpha = 0.00729735253594 = 1/137.03599971 \tag{17}$$

$$\frac{\mathfrak{R}_0}{\lambda_e} = 0.0073142871641 = 1/136.7187229$$

The Feynman Photon-Photon Electric Attraction, Repulsion and the Chirality of the Electron

D.T. Froedge

V042422

Abstract

Papers previously presented in regard to the interaction of charged particles have not addressed the issue of the $\pm$ value of charge $Q = \sqrt{\alpha c h}$. It is presumed that there is no substance that can be identified as charge. Charge is a regarded as a fictitious substance providing equal and opposite forces on particles defined in the framework of an Electric Field. The Electric Field is regarded as a very useful but also fictitious construct.

The interaction of masses by the interference of the Feynman photon can be shown to provide the equivalent interaction without the handicap of energy densities and infinities, as well at the interchange of imaginary particles.

In a previously presented paper: "The Electron as a Composition of Two Vacuum Polarization Confined Photons Revised"[5]. A model of an electron was presented that shows the electron to be a pair of photons each with half the energy of an electron, in a closely bound orbit, held together by the mutually induced index of refraction of the vacuum polarization. The rotating photon action paths for a pair of photons exist throughout space and provide the mysterious spooky action at a distance of the forces of fields.

For a circularly polarized, spinor defining photon momentum, rotating in a circle, the Electric vector is constantly in the radial direction, and the magnetic vector is aligned along the perpendicular angular momentum vector. For an opposite charge turning in the same direction the magnetic vectors aligned with the angular momentum are opposite. The electric and magnetic vectors are not considered as electric field stress vectors, but the probability flow direction of the location of the photon and its axial vector acceleration

From Feynman's action path considerations, the action path probability of these photons constitutes the angular momentum, the spin, and the anomalous spin. There is created a time independent probability of existence throughout space that can interact with the photons of other particles. The interactions of the probability existence of these photons, with the probability of photons from other particles, induce the effect of charge.

Particles have no charge and thus attraction and repulsion must occur as the result of the photon-photon rotational interaction of the photons.

The spin is the rotation of the central two particles and the probability paths throughout the surrounding space.

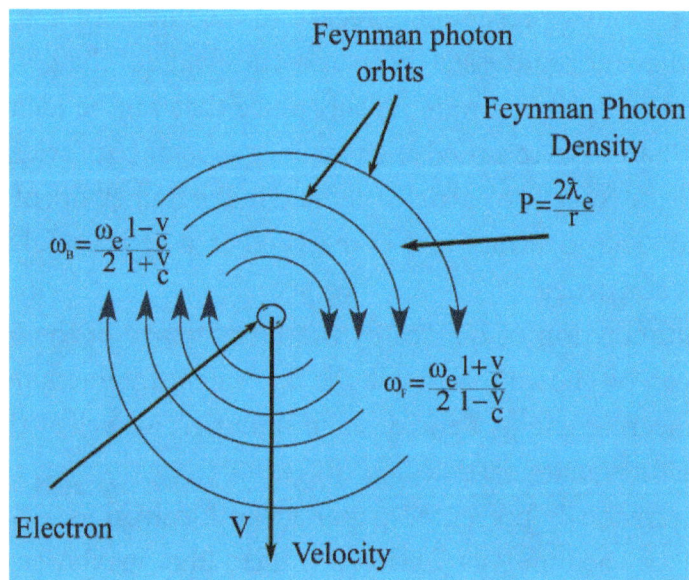

Fig. 1 The Electron showing Feynman action paths from a point perpendicular to the plane of rotation

$$\omega_B = \frac{\omega_e}{2}\frac{1-\frac{v}{c}}{1+\frac{v}{c}}$$

$$\omega_r = \frac{\omega_e}{2}\frac{1+\frac{v}{c}}{1-\frac{v}{c}}$$

Feynman photon orbits

Feynman Photon Density

$$P = \frac{2\lambdabar_e}{r}$$

Electron

V

Velocity

As the electron moves with spin perpendicular to the velocity Fig. 1, there is a relativistic Doppler difference in the frequencies observed on opposite sides of the spinning particle. The difference in these frequencies constitutes cause of both the deBroglie momentum-wavelength and the diffraction pattern of the Electron.

Charge

The change in c as the result of a charged particle is exactly as for gravitation accept that the photons are rotating around an axis remain polarized.

The effect of one charge on another is not as simple as it might at first. Unlike gravitation In which the exterior Feynman protons are generally random, in the case of electric effect, the relative direction of the density flow creates the effect of charge. Photons going in the same direction don't interact whereas photons going in opposite direction reduce the velocity of both flows in the region of the planar rotation, and attract the center of mass of each other.

It is not the purpose of this paper to calculate the exact value of the interaction which has been done elsewhere, but to show the mechanism of positive and negative charge occurs without the concept of charge.

Electron Angular Momentum

The angular momentum of the electron is the sum of the rotation of the two Feynman photons and the rotation around the center of mass, or orbital angular momentum.

$$\vec{J} = \vec{L} + \vec{A}_s \qquad (1.1)$$

As is known, the total angular momentum J, combines both the spin and orbital angular momentum. QM asserts that angular momentum applies to J, but not to L or S; spin–orbit interaction allows angular momentum to transfer back and forth between L and S, with the total remaining constant.

Also from standard QM the magnetic momentum vector $\bar{\mu}_J$ does not necessarily line up with the total angular momentum $\vec{J} = \vec{L} + \vec{A}_s$, but the Wigner-Eckart theorem [6], finds the expectation value does effectively lie on the direction of $\vec{J}$, thus $\bar{\mu}_J = \bar{\mu}_L + \bar{\mu}_S$. If, as the previous papers have demonstrated, $\bar{L}$ and $\bar{L}$ are just the same photon motion observed from different coordinate centers generated by the same photons then this makes logical sense.

The electron is a composition particle of two vacuum polarized photons, thus the sum of the angular momentum, for both the internal rotation and the orbital rotation for the two photons in, Eq. (1.1), is:

$$\vec{J} = \vec{R} \times (m_1 + m_2)\vec{v} + \sum_i r_i \times (m_1 + m_2)c \qquad (1.2)$$

or

$$\vec{J} = (m_1 + m_2)(\vec{R} \times \vec{v} + r_i \times \vec{c}) \qquad (1.3)$$

The two rotating photons of the electron [3], [5], can't simultaneously rotate about two axes and must have a single rotation axis, thus the **spin is the minimum of the total angular momentum of the electron,** and is a physical justification basis for the Wigner-Eckart theorem.

For the two photon composed electron **the spin angular momentum is not** separate, or independent, of its orbital angular momentum, but part of it. The corollary to this is that the velocity vector is always in the plane of the rotation, and the spin is perpendicular to the velocity.

The theory being proposed here requires the, velocity vector to be in the plane of the angular momentum and spin. If the rotation of the Feynman photons did not reside in the plane of the velocity, (Fig, 1), the deBroglie frequency and wavelength, as determined by the difference in the Doppler shift $p = \hbar\Delta\omega / c$ would be wrong, and the electron slit diffraction pattern would be a random smear.

To restate, the total angular momentum is the sum of the two rotations in along the same axis.

$$\vec{J} = \overrightarrow{L + A_s} \qquad (4)$$

Electron Spin and Angular Momentum

The angular momentum for rotation is defined as right handed, and the angular momentum for the Feynman photons for both type of charged particles must also be right handed. The electric vectors have the same frequency of rotation as they rotate around the electron axis of rotation thus the electric vectors are frozen along the radial direction. The difference in the electron and the positron is that for the same value of the angular momentum the magnetic vector is up for one and down for the other. The vectors designated as electric and magnetic do not have charge effect, but represent location flow and flow acceleration probability directions.

The spin vector as we know it as measured by its magnetic moment is the product of the angular moment and relative direction of the magnetic vector of the radially polarized Feynman photons. Two Feynman photons from opposite charge particles rotating in the same direction have opposite magnetic vectors, thus the spin vector of the charge particle is:

$$\vec{S}' = \pm \vec{A}_F \tag{5}$$

The $\pm$ term is just the sign of the conventional left or right hand chirality of the particle. Since the emitted electrons in the Weak Force experiment have been measured and designated as being left handed, it is the required convention that all the electrons in the universe be left handed, thus for the Electron with left hand chirality.

$$\vec{J}_e = \overrightarrow{L - S} \tag{1.6}$$

The spin vector designation is opposite the angular momentum. For the Positron with right hand chirality $\vec{J}_P$, is:

$$\vec{J}_p = \overrightarrow{L + S} \tag{1.7}$$

This sign convention provides the charge $\pm$ effect for the electron-positron interaction.

Opposite Charges

For a simple atomic system, Positronium, there are two opposite, rotating particles. Each particle is surrounded by a circulating probability of the core photons being in a planar rotating path. As shown in Fig. [2], the photons from both particles rotate, and engage the photons from the other particle.

Attraction or repulsion between the particles depends on direction of the interacting photons. This is the result of the Lorentz condition that photons in the same direction cannot interact, whereas opposite going photons have a probability of collision, and reduce each other's relative speed of light, or index of refraction.

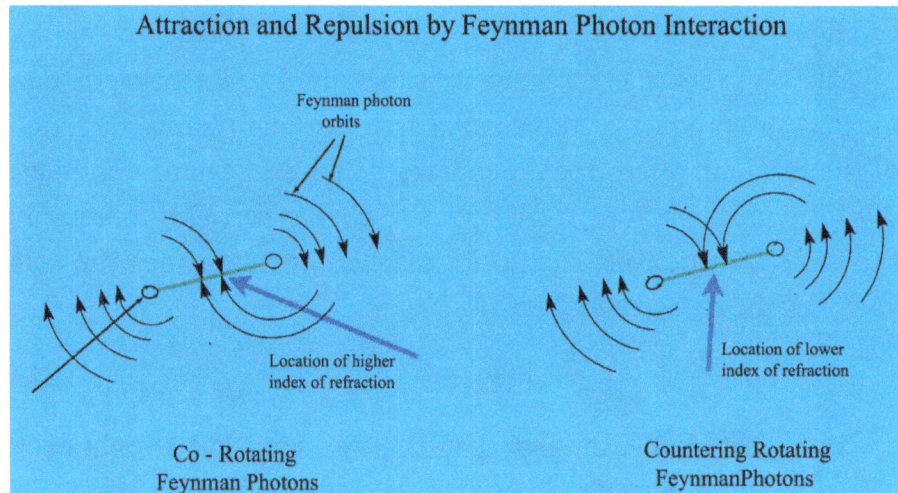

Attraction and Repulsion by Feynman Photon Interaction

Feynman photon orbits

Location of higher index of refraction

Location of lower index of refraction

Co - Rotating Feynman Photons

Countering Rotating FeynmanPhotons

Fig, 2. Interacting rotating particles

The direct collision probability illustrated in appendix I, shows the magnitude of the change in c as function of r, Eq.(1.14)

$$\frac{\Delta c}{c} = \frac{1}{2} \frac{1}{m_e c^2} \frac{\alpha c \hbar}{r} \tag{8}$$

Δc is the change in the velocity of light or related to the index of refraction as:

$$1 - \eta^{-1} = \Delta c \tag{1.9}$$

From Appendix II, the interaction for photons engaging the

60

probability flow at any angle Δc is a function of the angle of the interaction or:

$$\frac{\Delta c}{2c_0} = \left[\frac{\Delta c}{2c_0}\right]_{\perp} \sqrt{\frac{[1-\cos\theta]}{2}} \qquad (10)$$

The θ angle, is the angle between the interaction photons. From the equation photons moving in the same direction $(\cos\theta = 1)$, have no interaction.

Attraction

If the angular momentums of the particles are aligned, in the spherical volume between the rotating particles, the Feynman Photons interact. The interaction magnitude or the change in the index of refraction of two photon flow probabilities is maximum when the dot product is negative $(\hat{P}\cdot\hat{P} = -1)$, indicating a head on collision, and a decrease in the speed of light per Eq. (8)

$$\frac{\Delta c}{c} \sim \sqrt{\frac{[1-\cos\theta]}{2}} \qquad (11)$$

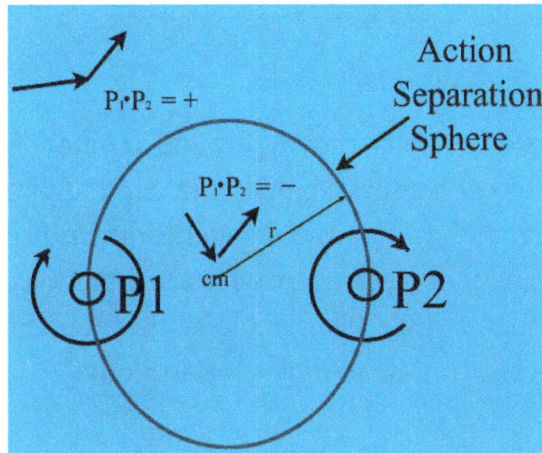

Fig, 3. The dot product of Feynman photons from the particles switches from positive to negative at the imaginary sphere between the co-rotating particles

The dot product is negative inside the sphere Fig. 3, and positive outside.

For the interacting particles along the line connecting the particles the increase in the gradient is in the direction of the opposite particle.

$$P = \frac{2\lambdabar_e}{r_1} = \frac{2\hbar}{m_e c r_1} \tag{1.12}$$

$$\frac{dP}{dr} = -\frac{2\lambdabar_e}{r_1^2} \tag{1.13}$$

This increasing gradient in c or the index of refraction causes the Feynman action path of the photons of the second particle to move in the direction of the other particle, and causing the orbiting planes to align.

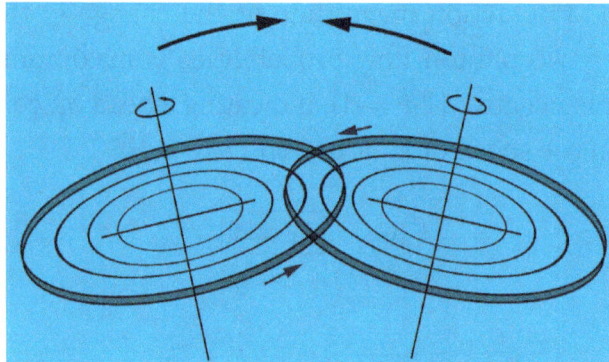

Fig. 4. The rotating planes of the Feynman photon. Represent a reduction in the in the spatial index of refract and induce the planes to alignment

As a photon probability moving around a center of mass on circular action paths encounters a gradient in c in the direction of a second particle the path will be diverted in the direction of that particle, and thus the center of mass is moved in the direction of the center of mass of the other particle. This would be the interaction of opposite charged particles, and the particles are attractive.

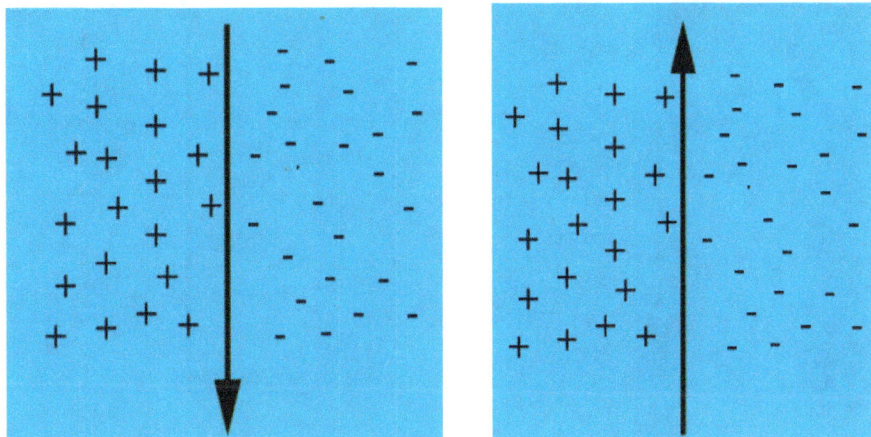

Attraction -Spins Anti-Aligned
Photons probabilities in opposite direction here
V

Electron
Left Hand Chirality

Positron
Right Hand Chirality

Fig. 5. Feynman probability photon rotation around the core photons Positive (+), Photons out of page. Negative (-) Photon into page.

Repulsion

If the particles have counter-rotating photons then at the same location in that sphere along that line the photons are going in the same direction with a dot product of +1 and do not interact. The maximum value of the dot product for this interaction maximizes on the opposite side of the opposite particle outside the sphere, pulling the particles apart.

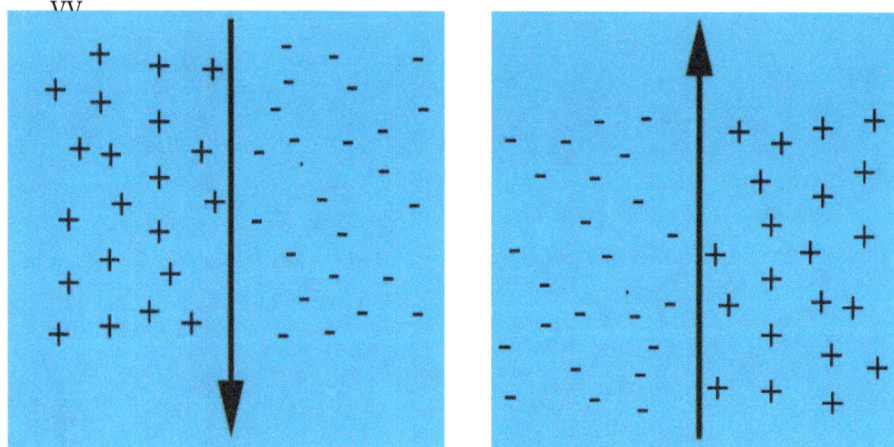

Fig. 6. Feynman probability photon rotation around the core photons Positive (+), Photons out of page. Negative (-) Photon into page.

Electron
Right Hand Chirality

Electron
Right Hand Chirality

Critical to the process of charge is the chirality of the electron and the positron. It is normally assumed that the Electron and Positron can have spin up or down independent of their angular momentum or their charge. This development doesn't suggest that all, it presumes that the spin is the value of the angular momentum. The vector sum of the spin and orbital angular momentum is conserved, and thus project onto the same axis.

The procedure outlined here provides the mechanism of the origin of the ± sign of the charge.

The Pauli Exclusion Principle results from the fact that in an atomic system the spins of alike particles anti-align in pairs for minimum interference,

Constraints on Solutions to the Dirac Equation

Attraction and repulsion of Electrons and Positrons by the interaction of Feynman photons as defined here, depends on the chirality of the particles, and it's not optional. Electrons and positrons have to have opposite chirality in order to attract and repel. If the selection were optional, electrons would attract or repel other electrons or positrons.

The emitted electrons in the Weak Force experiment have been

64

measured and designated as being left handed, thus consistent designation for all electrons is to have left hand chirality, and right hand chirality for positrons..

This selection puts limits on the solutions to the Dirac equation. There are four solutions to the Dirac equation, two for an Electron and two for a Positron, each with left or right chirality [7].

The requirement here thus eliminates two solutions of the Dirac equation as being inoperative in the real world.

The other solutions may exist, and may be creatable in accelerators or particle reaction, but those particles cannot exist in nature. This consideration eliminates parity violation in Beta decay.

Appendix I

Dependence of Photon Direction on Photon-Photon Interactions

$$\frac{\Delta c}{2c} = \frac{1}{m_e c^2} \frac{Q^2}{r} = \frac{\Delta \varepsilon}{\varepsilon_T} \tag{1.14}$$

Δc, is the change in the index of refraction for a photon moving opposite the Feynman photon density of an electron at a distance r.

From [3], the wavefunctions derived momentum 4-vector of two equal photons is:

$$\vec{P_1} = \frac{\hbar \omega_1}{c^2} \left(\gamma^k c_{1k} + \gamma^0 c \right) \qquad \left(\gamma^k \right)^2 = -1 \qquad \left(\gamma^0 \right)^2 = 1 \tag{1.15}$$

$$\vec{P_2} = \frac{\hbar \omega_2}{c^2} \left(\gamma^k c_{2k} + \gamma^0 c \right)$$

The scalar invariant sum of the momentum is:

$$E_I = c \left| \vec{P_1} + \vec{P_2} \right| \tag{1.16}$$

Presuming the photon trajectories are defined in the same plane

$$E_I = c\left(\vec{P_1} + \vec{P_2}\right)\left(\vec{P_1} + \vec{P_2}\right)^* = c\left(\vec{P_1}^2 + \vec{P_2}^2 + \vec{P_1}\bullet\vec{P^*_2} + \vec{P_1}\wedge\vec{P^*_2} + \vec{P_1}\bullet\vec{P^*_2}\right)$$

$$\vec{P_1}^2 = 0 \qquad \vec{P_2}^2 = 0 \qquad \vec{P_1}\wedge\vec{P_2} - \vec{P_2}\wedge\vec{P_1} = 0 \tag{1.17}$$

Thus:

$$E_I = +2\vec{P_1}\bullet\vec{P_2} \tag{1.18}$$

Specifically from [3], this is:

$$2\vec{P_1}\bullet\vec{P^*_2} = \left[-\frac{\hbar\omega_1}{c^2}\gamma^k c_{1k}\bullet\frac{\hbar\omega_2}{c^2}\gamma^k c_{2k} \quad +1\right] \tag{1.19}$$

Or:

$$E_I = E_T\sqrt{\frac{[1-\cos\theta]}{2}}$$

Thus

In the case of the interaction of Feynman photons his energy is a Lorentz scalar energy or rest frame energy associated with the slow-down of the photons E_I, thus

$$\frac{\Delta c}{2c} = \left[\frac{\Delta c}{2c}\right]_0\sqrt{\frac{[1-\cos\theta]}{2}} \tag{1.20}$$

And if $\cos\theta = 1$, indicating the angle between the direction of photon is zero the photons do not interact.

References:

Preceding Papers by Author

1. DT Froedge, The Connection between Electric Charge, Gravitation, and the Feynman Sum over All Histories View of Quantum Electrodynamics, April 2020 Conference: APS April. 18-21, 2020 Washington DC.
https://absuploads.aps.org/presentation.cfm?pid=18355
https://www.researchgate.net/publication/341310206

2. DT Froedge, A Quantum Theory Conjecture on the Origin of

Gravitational and Electric Particle Interaction, December 2019,
DOI: 10.13140/RG.2.2.29097.54884
https://www.researchgate.net/publication/337826826

3. DT Froedge, The Dirac Equation and the two Photon Model of
the Electron
February 2021, https://www.researchgate.net/
publication/349089256

4. DT Froedge, The Gravitational Constant to Eleven Significant
Digits,
March 2020, DOI: 10.13140/RG.2.2.32159.38564
https://www.researchgate.net/publication/339943651

5. DT Froedge, The Electron as a Composition of Two Vacuum
Polarization Confined Photons Revised Apr. 2021 DOI:
10.13140/RG.2.2.18971.18722 https://www.researchgate.net/
publication/350740864

Other Papers

6. K.T. Hecht (2000) The Wigner—Eckart Theorem. In: Quantum
Mechanics. Graduate Texts in Contemporary Physics. Springer,
New York, NY. https://doi.org/10.1007/978-1-4612-1272-0_32

7. R. Hong, et. al., Helicity and nuclear βdecay correlations, Am.
J. Phys. 85, 45 (2017); doi: 10.1119/1.4966197 http://dx.doi.
org/10.1119/1.4966197

8. DT Froedge, The Flaw in Quantum Mechanics revised
January 2020, DOI: 10.13140/RG.2.2.18520.44804

8. S. Errede, The Structure of Space-Time, 2015Lecture Notes 16,
EM Fields & Sources II FallSemester,http://web.hep.uiuc.edu/home/
serrede/P436/Lecture_Notes/P436_Lect_16.pdf

The Connection Between Electric Charge, Gravitation, and the Feynman Sum Over All Histories View of Quantum Electrodynamics

D.T. Froedge

V041120

Abstract-

In the introduction to Feynman's *Six Easy Pieces* it was said: "You could not imagine the sum-over-histories picture being true for a part of nature and untrue for another part. You could not imagine it being true for electrons and untrue for gravity" The purpose of this paper is to show that Gravitation and Electric Charge are the result of the interaction of the Feynman path photons. Feynman proposed that for a photon, or any particle, going from one point to another[1], there is a probability of the particle has traveled every possible path, and by very accurate measurements of quantum effects there is every reason to believe that this is true. It is shown that the interaction of these Feynman path photons generated by mass particles change the index of refraction of space, and can constitutes the effects of gravitation and electric charge.

It is proposed that a photon moving through probability density amplitude of approaching Feynman photons experiences an alteration in the index of refraction. This alteration of photon dynamics can be shown to be the causation of gravitation, and, electric charge

It is remarkable from this relationship that the value of the **Gravitational Constant can be calculated** to at least **eleven significant digits.** It is further remarkable that the value is **within the error bars** of all the measurements published by the by the **International Bureau of Weights and Measures, (BIPM),** since 2000 (Appendix I)

This paper is compiled from several of the authors papers, [2],[3],[4],[5].

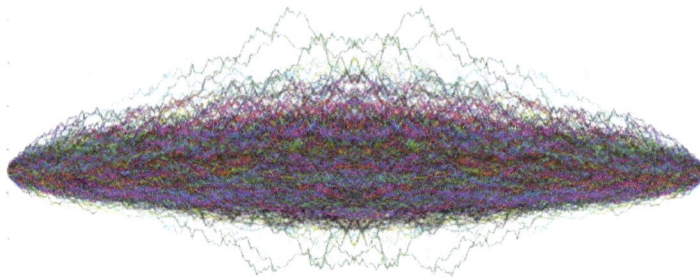

Fig.1, A subset of the infinite set of trajectories

For photons on repetitious tracts such as a standing wave or oscillation in a particle, there is a continuous repetition of an infinite number of external paths.

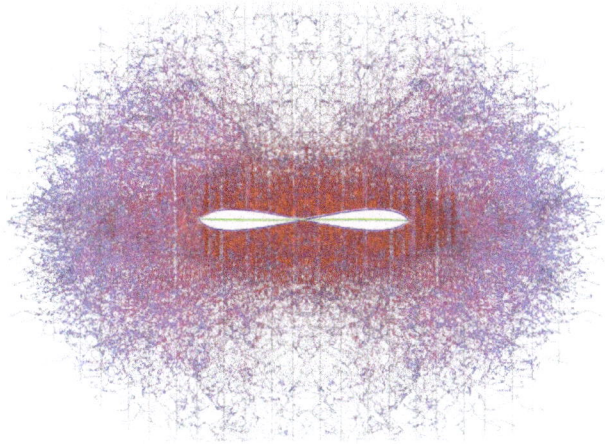

Fig. 2 The probable path density, which is infinite is not the same as the location probability density.

Although the probability of a Feynman action path for a photon can be anywhere, the probability is not infinite everywhere, but is a function of the distance form the most probable path Fig, 2.

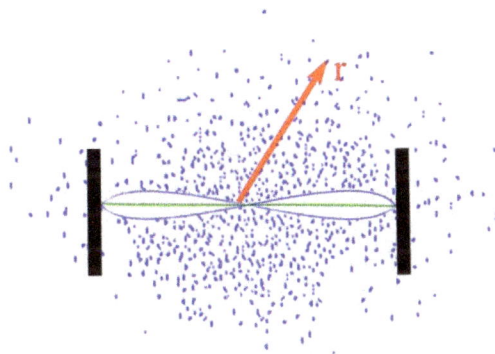

Fig. 2, probability of Feynman photons exterior to the minimum path

The probability density of the photons form the most probable paths has been investigated by a number of researchers and summarized in[4]

$$P \sim 1/r \tag{1}$$

Sakoda and Omote [6] and others found the scattering amplitudes finding the asymptotic probability distribution $r \gg \lambda$ to be proportional to:

$$\mathbf{P}(r_\perp) = \psi^*\psi \sim \frac{\lambda}{r} \tag{1.2}$$

For the consideration of a trapped photon oscillating between two points or in a bound repetitious motion Fig. 2 , the time average probability density of the Feynman photons being at a position at a distance r from the classical tract is found to be:

$$P_F = \frac{2\lambda_C}{r} \tag{3}$$

P is the probability density λ_c is the Compton wavelength and r is the radial distance from the minimum path.

Photons orbiting inside mass particles are not necessarily axially aligned with other particles and do not have a common rotation axis thus this expression is the time average density of all random directed photons and constitutes the probability density of photons generated by a mass. Fig.4

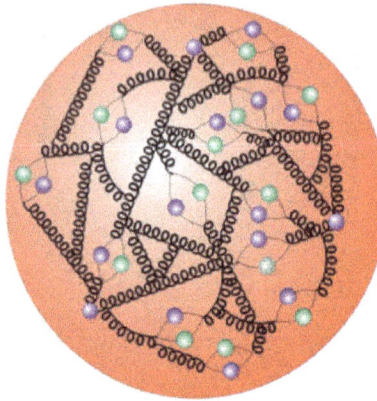

Fig. 4, Illustration of random constituents is a mass particle including gluons bosons leptons quarks etc.

The presumption is made that at the subatomic level the mass of any particle is composed of bound light speed of particle that travel repetitive paths and producing action paths that exist outside the particle. These particles are responsible for producing both gravitational and electrical interactions it is presumed that this probability density is a general property of matter and applies to all mass particles.

Postulate Regarding Feynman photons interact by altering the velocity of light

Postulate: The change in the speed of light, on passing through the Compton volume of a second photon is proportional to the ratio of the Planck particle area $\left(\lambdabar_{PL}^2\right)$ to the Compton area $\lambdabar_c^2$

$$\frac{\Delta c}{c_0} = \frac{\lambdabar_{PL}^2}{\lambdabar_c^2} \tag{1.4}$$

The related index of refraction, η, and Planck constant $\lambdabar_{PL}$ used here are:

$$\frac{\Delta c}{c_0} = \frac{c_0 - c}{c_0} = \left(1 - \frac{c}{c_0}\right) = \left(1 - \eta^{-1}\right) \qquad \lambdabar_{PL} = \sqrt{\mu\lambda} = \sqrt{\frac{G\hbar}{c^3}} \tag{5}$$

Eq. (1.4), presumes the photon is the size of the Planck particle its probability density located within the Compton radius and the change in c is just the probability of a direct hit of the Planck particle. (Fig. 5)

Change in c for Photon Resulting from Probability
of Intersecting Planck core of 2nd Photon in its
Location Volume

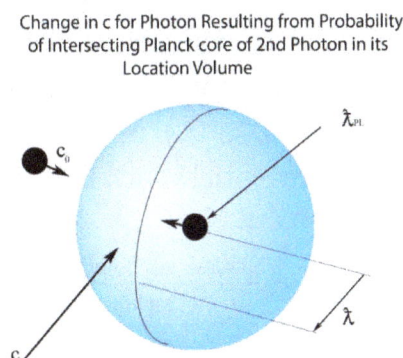

Fig.5 , Photon delay passing through probability distribution of another photon.

Multiplying the change in the speed of light per photon times the probability density of Feynman photons, Eq.(3), Eq.(1.4), gives:

$$\frac{\Delta c}{c_0} = \text{Change per photon} \times \text{photon probability denstiy}$$

$$\frac{\Delta c}{c_0} = \frac{\lambdabar_{PL}^2}{\lambdabar_c^2} \times \frac{2\lambdabar_c}{r} \tag{6}$$

Or

$$\frac{\Delta c}{c_0} = \frac{\text{Target area}}{\text{Total area}} \text{ density}$$

This fundamental relation which defines the change in the velocity of light induced by Feynman photons is the source of both gravitational and electric interaction. This has connection to the work of M. Urban et.al.in regard to the vacuum background [7],[8].

From Eq.(6), this immediately gives the speed of light in the presence of mass to be the gravitational relation.

$$\frac{\Delta c}{c_0} = \frac{2\mu}{r} \tag{7}$$

or

$$c = c_0 \left(1 - \frac{2\mu}{r} \right) \tag{8}$$

This is the exact change in the index of refraction of light in the presence of a mass as predicted by General Relativity when projected on flat Minkowski space, and confirmed by the radar measurements of the Shapiro effect. [9],[10],[11].

Restating in terms of potential which is taken as universal:

$$\frac{\Delta c}{2c_0} = \frac{GM}{c^2 r} = \frac{GMm_1}{(m_1 c^2) r} = \frac{\text{Gravitational potential energy}}{\text{Total energy}} \text{ per unit mass}$$

$$\frac{\Delta c}{2c_0} = \frac{\Delta E}{E} \tag{9}$$

The velocity ratio term in Eq. (9), $\Delta c / 2c_0$ is equal to the ratio of the potential to total energy of interacting particles, and the same for both electric and gravitational interactions. In a conservative system it is just equal to the ratio of the kinetic to rest energy.

Electric Interaction

Using the same postulates and presumptions on the Feynman the electric charge interaction can be found.

The difference between gravitation and electric interaction is that the probability of Feynman photons from an electron does not move in random directions, but due to spin polarization, move as the core particles of the electron in a circular path around the particle [2].

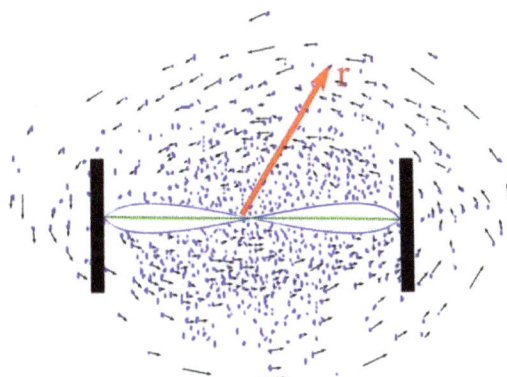

Fig. 5, Circular directions of Feynman Photons for polarized charged mass particles

This constitutes an increase in probability density for an oncoming photon by a factor of the repetition rate. An oncoming photon experiences the static density Eq. (10), times the repetition rate of that value (Fig.4 & Fig.5).

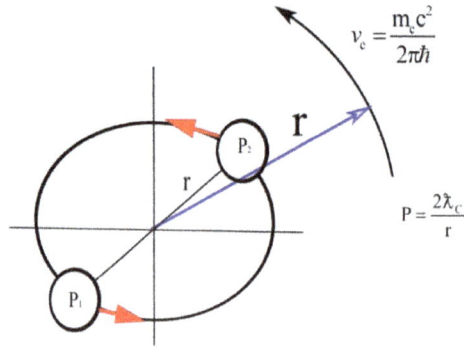

Fig. 4, Shows the time average probability density for a Feynman photon on an action path and the repetition frequency an any point.

This means that at each cycle of the internal photons at the Compton frequency there is a path repetition making another contribution to the probability density as illustrated in Fig. 7:

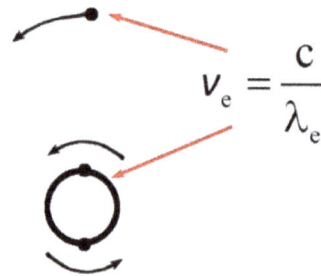

Fig. 7, Frequency of probable photon passing a point equal to cycle of the internal photons.

Thus the probability density of the location of a Feynman photon interacting with an oncoming photon from another electrical charge Eq. (3), is then multiplied by the frequency:

$$P_F = \frac{2\lambda_C}{r} \, \nu_e \tag{10}$$

ν_e is the Compton frequency of the electron, and the expression for the change in the speed of light in the presence of an electron (Eq. (6)) becomes:

$$\frac{\Delta c}{c_0} = \text{Change per photon} \times \left[\text{Photon probability denstiy} \times \text{Frequencyp of path repetition} \right]$$

76

$$\frac{\Delta c}{c_0} = \left(\frac{\bar{\lambda}_{PL}^2}{\bar{\lambda}_c^2} \times \frac{2\bar{\lambda}_C}{r} \times v \right) \tag{11}$$

Compare this with the expression for Gravitation Eq.(6),

$$\frac{\Delta c}{c_0} = \frac{\bar{\lambda}_{PL}^2}{\bar{\lambda}_c^2} \times \frac{2\bar{\lambda}_C}{r}$$

This is the change in c as the result of the electron photon probability density, but the interaction of the photons between electrons requires the probability of the coincidence thus the product of the probabilities. It is the interaction between the particles that determines the coupling constant for the electrons (see Fine Structure Constant Appendix I).

Multiplying interactions of two particles in proximity gives:

$$\frac{\Delta c}{2c_0} = \left(\frac{\sqrt{2}\bar{\lambda}_{PL}^2}{\bar{\lambda}_c^2} \frac{c}{\lambda_e} \right)\left(\frac{\sqrt{2}\bar{\lambda}_{PL}^2}{\bar{\lambda}_c^2} \frac{c}{\lambda_e} \right)\frac{\lambda_e}{r_1} \frac{\lambda_e}{r_2}$$

For this to be correct as the change in the velocity of light in the presence of two electrons, the quantities in parenthesis must be the fine structure constant α thus:

$$\frac{\Delta c}{2c_0} = \frac{\alpha\lambda_e}{r_1} \frac{\alpha\lambda_e}{r_2} \tag{12}$$

The value of alpha α can be then written as:

$$\alpha = \frac{\sqrt{2}\,\bar{\lambda}_{PL}}{\bar{\lambda}_e} v_e \;\rightarrow\; \alpha = v_e\sqrt{\frac{2\mu_e}{\bar{\lambda}_c}} \;\rightarrow\; v_e m_e \sqrt{\frac{2G}{c\hbar}} \tag{13}$$

$$\alpha = 0.00729735253596$$

Appendix I shows proper inclusion of the gyromagnetic ratio and the value of G in the calculations to obtain accurate values. Only the error in the experimental value of G limits the accuracy of the calculation. By calculating the value of G from the other constants its precise value can be found to at least eleven significant digits.

The value of the index of refraction in this expression, Eq.(12), is at the position r_1 from the first particle thus valuating the index of refraction at the classical radius $r_1 = \alpha\lambda_e$ of the first particle gives, the value function of the distance to the second particle. This is then the ratio of the potential energy to the total to the total energy of the particle, or the energy ratio ot the potential to the total energy of the electron.

$$\frac{\Delta c_I}{2c} = \frac{1}{m_e c^2} \frac{Q^2}{\Delta r_{12}}$$ (1.1)

The interaction of two mass particles by gravitation

The gravitational interaction of two masses can be found using the same procedure for gravitation as charge, and Eq. (7).

The combined effect of the two particles at a point at a distance r from the fist point and that point being Δr_{12} from that point is:

$$\frac{\Delta c}{c_0} = \frac{\Delta c_1}{c_0} \frac{\Delta c_2}{c_0},$$ (1.2)

The total index of refraction at change at is then:

$$\frac{\Delta c_T}{c_0} = \frac{2\mu_1}{r_1} \frac{2\mu_2}{\Delta r_{21}}$$ (1.3)

Since the value of the index of refraction is valid at any point it is convenient to pick a point a constant relative point independent of the relative motion

For the same reason as for the electric evaluation the evaluation point is set at the sum of the Schwarzschild radius.

$$|\vec{r}| = (2\mu_1 + 2\mu_2)$$ (1.4)

then

$$\frac{\Delta c}{c_0} = \frac{2\mu_1}{2(\mu_1 + \mu_2)} \frac{2\mu_2}{\Delta r_{12}}$$ (1.5)

Writing this out explicitly gives the index of refraction in terms of the particle separation thus:

$$\frac{\Delta c}{2c_0} = \frac{1}{(m_1 + m_2)c^2} \frac{Gm_1m_2}{\Delta r_{12}}$$

(1.6)

This is the proper value of the ratio of the potential energy to total energy of two gravitating particles or the energy extractable by a change in relative position.

Energy and the Index of Refraction

The first point is to note that the gravitation and electric potentials for an electron in Eq.(1.6), and Eq.(1.1), are identical accept for the factor of the Compton frequency ν

$$\text{Gravitation} \qquad \frac{\Delta c_e}{c} = \frac{\mu_e}{\lambda_e} \frac{2\lambda_e}{r_1}$$

(1.7)

$$\text{Charge} \qquad \frac{\Delta c_e}{c} = \frac{\mu_e}{\lambda_e} \frac{2\lambda_e}{r_1} \nu_e$$

(1.8)

The Δc term appears a factor in both potentials and is related to:

$$\frac{\Delta c}{2c_0} = \frac{1}{mc^2} \frac{Q^2}{\Delta r_{12}} = \frac{\varepsilon_P}{\varepsilon}$$

$$\frac{\Delta c}{2c_0} = \frac{1}{(m_1 + m_2)c^2} \frac{Gm_1m_2}{\Delta r_{12}} = \frac{\varepsilon_P}{\varepsilon}$$

(1.9)

The left side of all these expressions is the ratio of the extractable potential energy for two particles to the total energy of the mass.

The relativistic change in energy to the total energy can be expressed the same form.

$$\frac{\Delta c}{2c_0} = \frac{(m_0 - m)c^2}{m^2c^2} = -\frac{1}{2}\frac{v^2}{c^2} = \frac{\Delta \varepsilon_K}{\varepsilon}$$

(10)

Then Eq,(9).and Eq.(10), allows the formulation of the Lagrangian for both Gravitation and charge as:

$$1 - \frac{c_0 c}{c_0^2} - \frac{v^2}{c_0^2} = \text{Constant} \qquad (11)$$

Summarizing, it can be seen that the half ratio of the change in the speed of light to the ambient velocity by the Feynman photon density is equal to the ratio of the potential energy to the total energy of the mass of interacting particles. This is true for both charge and gravitation interactions.

Conclusion

Presented has been a plausible causation of gravitation and electrical charge interaction within the confines Quantum Theory in Minkowski four-space. The connection between the Gravitational constant and the Fine Structure constant presented here is unprecedented.

It has to be regarded as a new approach to physics, and as such, a degree of speculation has been incorporated, many parts lack mathematical rigor, but fit well with known physical parameters. How the postulates and speculation impact or elucidate the understanding of QM will be left to future clarification.

Hopefully the ideas presented here will lead to a better understand between Quantum Mechanics and Gravitation

References:

1. Feynman, Hibbs, 1965, Quantum Mechanics and Path Integrals McGraw-Hill

2. DT. Froedge, The Electron as a Composition of Two Vacuum Polarization Confined Photons Revised, February 2020, DOI: 10.13140/RG.2.2.31942.22085, https://www.researchgate.net/publication/339512823

3 DT. Froedge, A Quantum Theory Conjecture on the Origin of Gravitational and Electric Particle Interaction, December 2019, DOI: 10.13140/RG.2.2.29097.54884, https://www.researchgate.net/publication/337826826

4. DT. Froedge, Quantum Field Origin of Gravitation, APS, April, 2019; Denver
http://meetings.aps.org/Meeting/APR19/Session/H11.6

5. DT. Froedge, The Gravitational Constant to Eleven Significant Digits, March 2020, https://www.researchgate.net/publication/339943651

6. S Sakoda , M Omote, Difference in the Aharonov-Bohm Effect on Scattering States and Bound States Difference in the Aharonov-Bohm Effect on Scattering, Advances in Imaging and Electron Physics, v 110, 1999, Pages 101-171, Academic Press, 1999

7. M. Urban, F. Couchot, X. Sarazin, A. Djannati-Atai, The quantum vacuum as the origin of the speed of light, EPJ manuscript, arXiv:1302.6165v1 [physics.gen-ph] 2013

8. M. Urban A particle mechanism for the index of refraction, LAL 07-79, 2007,
Dmitri Kharzeeva and Kirill Tuchinb,Vacuum self–focussing of very intense laser beams, BNLNT-06/43, RBRC-657,arXiv:hep-ph/0611133v2 21 Feb 2007 https://arxiv.org/abs/0709.1550

9. Roger Blandford, Kip S. Thorne, Applications of Classical Physics, (in preparation, 2004), Chapter 26
http://pmaweb.caltech.edu/Courses/ph136/yr2012/1227.1.K.pdf

10. J Chandler,et.al Solar-system dynamics and tests of general relativity with planetary laser ranging Dec. 2004, https://www.researchgate.net/publication/228831885

11. F. Karimi, S. Khorasani, Ray-tracing and Interferometry in Schwarzschild Geometry, arXiv:1001.2177
[gr-qc] arXiv:1206.1947v1 [gr-qc] 9 Jun 2012

12 G. Gabrielse et al,, New Determination of the Fine Structure Constant from the Electron g Value and QED,
Phys. Rev. Lett. 97, 030802 (2006)
http://hussle.harvard.edu/~gabrielse/gabrielse/papers/2006/
NewFineStructureConstant.pdf

13, G. Gabrielse et al Cavity Control of a Single-Electron Quantum Cyclotron Measuring the Electron Magnetic Momen,
arXiv:1009.4831v1 [physics.atom-ph] 24 Sep 2010

14 T. Quinn et.al, The BIPM measurements of the Newtonian constant of gravitation, G
https://royalsocietypublishing.org/doi/pdf/10.1098/rsta.2014.0032

15. T. Quinn,1, BIPM, Improved Determination Of G Using Two Methods, Physical Review Letter Sep 2013,
https://www.bipm.org/utils/en/pdf/PhysRevLett.111.101102.pdf

15. T. Quinn - A new determination of G using two methods. – NCB 2001 Phys Rev Lett. Sep 2001. Epub 2001 Aug.
27, https://journals.aps.org/prl/abstract/10.1103/
PhysRevLett.87.111101

18 C. Merkatas et. al.,Consensus Value for the Newtonian Constant of Gravitation
Xiv:1905.09551v1 [physics.data-an] 23 May 2019

19. C. Merkatas et.al. Shades of Dark Uncertainty and Consensus Value for the Newtonian Constant of Gravitation,
arXiv:1905.09551v1 [physics.data-an] , https://arxiv.org/
pdf/1905.09551.pdf

Appendix I

The calculation of the relation between the gravitational constant and the fine structurre Constant to eleven significant sigits

The interaction of the Feynman path photons resulting from the internal particle paths defines both gravitational and electrical forces [1],[2],[3]. As a result, the value of the fine structure constant is found expressed in terms of fundamental constants as:

$$\alpha^2 = \frac{2\lambdabar_{PL}^2}{\lambdabar_e^2} v_e^2 \tag{12}$$

The constants are the Planck radius, $\lambdabar_{PL} = \sqrt{\mu\lambda} = \sqrt{G\hbar/c^3}$, the Compton radius $\lambdabar_e$, the anomalous gyromagnetic ratio $g_A = g_e/2$, and the electron Compton frequency $v_e \to m_e c^2/2\pi\hbar$. The Feynman photons of two interacting particles represent the first order quantum loop, and thus the wavelength and the reciprocal of the frequency of the electron must be corrected by the higher order perturbations included in the anomalous gyromagnetic ratio, $g_A = g_e/2$, and $v_e \to v_e/g_A$. The values of the constants are included in Table [1].

All of the constants in the expression are known to at least eleven significant digits, except the gravitational constant. The maximum number of digits for the gravitational constant known with experimental certainty is about three.

By solving for the gravitational constant (G) in Eq.(1), the expression becomes:

$$G = \frac{\alpha^2 2\pi^2 c (\lambdabar_e g_A)^4}{\hbar} \tag{13}$$

From these values the gravitational constant can be calculated to an accuracy of about eleven significant digits:

G = 6.6755053318 x 10-8 cm³gm⁻¹s⁻²

This value is slightly higher (0.02%) than the current Codata consensus recommended value of 6.67430(15) x10- 11m3kg-1s-2, but it is not outside the scatter of measurements used in forming that consensus. It is within the error bars of all the experimental measurements conducted by the International Bureau of Weights and Measures, (BIPM), published since 2000, [4][5][6[[7]. Comparison with those values is presented below, Table [2].

The merit of this theory requires this relation to be correct. If it is wrong the theory is flawed.

Table 1

$c = 2.9979245800E+10*$ $g_A = g_e/2 = 1.00115965218$ †

$h = 1.054571817646E-27 *$ $\lambda_e = \hbar / m_e c = 3.8616633678E-11$

$a = 1/137.035999084$ † $m_e = 9.1093837015(28)E-28$ ⊕

∗ Definition† Gabrielse et. al. [12][13]. ⊕ 2018 Codata Recommended Values

Table 2

Comparing calculated value of G with BIPM and Codata consensus values

Calculated value	$G = 6.6755053318 \times 10^{-11}$
BIPM weighted mean 2014	$G = 6.67554(16) \times 10^{-11}$
BIPM Sep, 2015	$G = 6.67545(18)) \times 10^{-11}$
BIPM 01-32-2001	$G = 6.67559(27) \times 10^{-11}$
Codata Consensus Value	$G = 6.67430(15) \times 10^{-11}$

[14],[15],[16],[17]18],[19].

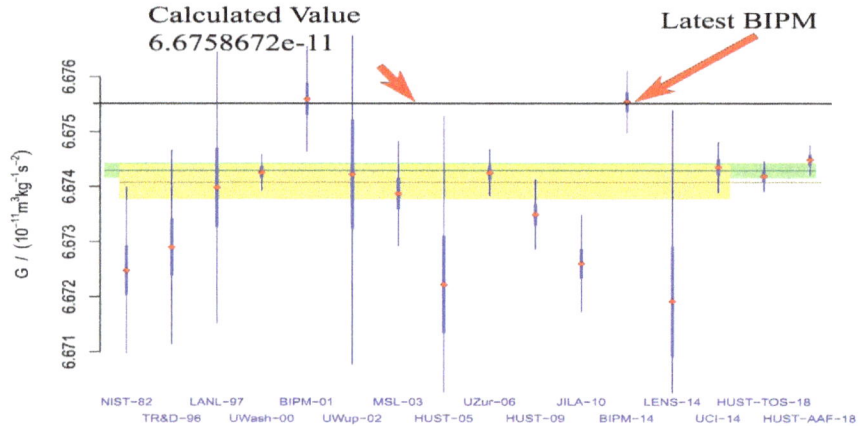

Experimental values used for statistical calculation of Codata consensus value of G, [13]

The Fine Structure Constant from the Feynman Path Integrals

Supplement to:

The Connection between Electric Charge,

Gravitation, and the Feynman Sum over All

Histories View of Quantum Electrodynamics

[1, 4]

D.T. Froedge

V031920

The referenced paper details this development, but this supplement is extracted from separate sections.

The change in the speed of c as the result of gravitation is well known. It is found from GR by flattening GR onto Minkowski four-space, and it is known by the experimental measurements of the Shapiro effect [6], [7], its value is:

$$1 - \eta^{-1} = \frac{\Delta c}{c_0} = \frac{2\mu}{r} \tag{1}$$

The relation between Δc and the conventional index of refraction is $1 - \eta^{-1} = \Delta c$. This same value can be calculated by the presumption that mass particles generate an extensive probability field of circulating photons as the result of the Feynman action path probabilities of the internal motion of the internal particles.

By presuming the change in c is proportional to the probability of a photon intersecting a photon the size of a Planck particle inside the volume of the Compton radius of a second photon, the change is found to be:

$$\frac{\Delta c}{c_0} = \frac{\lambdabar_{PL}^2}{\lambdabar^2} \tag{2}$$

$\lambdabar_{PL}$ is the radius of the Planck particle $\mu\lambdabar$, and $\lambdabar$ is the Compton radius of a mass particle. A probability density per unit volume of Feynman photons as a function of the distance from the particle generated by a mass particle has been found from other considerations to be [1]:

$$d = \frac{2\lambdabar}{r} \tag{3}$$

The radius r is the distance from an observation point of the density to the particle center of mass

The gravitational change in the speed of light in the vicinity of a mass particle is the probability of the change due to a photon times the probability density of encountering a Feynman photon from the mass particle in the vicinity of the particle:

$$\frac{\Delta c}{c_0} = \frac{\lambdabar_{PL}^2}{\lambdabar^2} \frac{2\lambdabar}{r} = \frac{2\mu}{r} \tag{4}$$

The same as the value as noted in Eq. (1). We take this as the origin of the change in c as a result of the Feynman photons generated by the internal action paths of photons in mass particles.

Rewriting this in in terms of the energy ratio of the gravitational energy to the total energy of the particle in the field gives:

$$\frac{\Delta c}{2c} = \frac{\mu}{r} = \frac{Gm_1}{c^2 r} = \frac{Gm_1 m_2}{(m_2 c^2) r} \tag{5}$$

To state this: It represents the ratio of the change in the potential energy, to the total energy of the m_2 particle, as the result of being in the probability density field of the Feynman photons of m_1

For a particle the change in the energy potential as a ratio of its total energy is then:

$$\frac{\Delta \varepsilon}{\varepsilon_T} = \frac{1}{2} \frac{\Delta c}{c} \tag{6}$$

Putting Eq.(4), into this is the form is then:

$$\frac{\Delta c}{c_0} = \frac{\lambda_{PL}^2}{\lambda^2} \frac{2\lambda}{r} \rightarrow \frac{\Delta \varepsilon}{\varepsilon_T} = \frac{1}{2}\left(\frac{\lambda_{PL}^2}{\lambda^2} \frac{2\lambda}{r}\right) \tag{7}$$

The Electron

An electron as modeled in [3], and [5], is defined as two photons captured in the self-generated index of refraction of the vacuum polarization. The modification to Eq.(7), which is the single pass of a photon through a probability field is that the photon probability from the rotation photons in the electron repeats at the Compton frequency, v_e and generates the effect of a higher change in the index of refraction on a passing photon.

$$\frac{\Delta c}{c} = \frac{\lambda_{PL}^2}{\lambda_e^2} \frac{2\lambda}{r} v_e = \frac{2\mu}{r} v_e \tag{8}$$

Note from Eq.(4), and Eq.(8), the effect of the Feynman photon density is the originating mechanism for both gravitation and electric charge.

In the case of electrons the change in c of one particle on the other has a reciprocity effect and the interaction on the change in c is multiplicative (see [1]). The relative change in c and the potential energy potential is the product of the individual changes thus:

$$\frac{\Delta c}{2c} = \left(\frac{\sqrt{2}\,\lambda_{PL}}{\lambda_e^2} v_e \frac{\lambda_1}{r_1}\right)\left(\frac{\sqrt{2}\,\lambda_{PL}}{\lambda_e} v_e \frac{\lambda_1}{r_2}\right) \rightarrow \frac{\Delta \varepsilon}{mc^2} = \frac{\alpha \lambda_e}{r_1} \frac{\alpha \lambda_e}{r_2} \tag{9}$$

Two values of r, are the distance of each paricle to an arbitrary observation point

$$\alpha = \frac{\sqrt{2}\,\lambda_{PL}}{\lambda_e^2}\,v_e \qquad (10)$$

The symmetric factoring allows the defining the interaction to be cast it terms of two equal and opposite charges and the energy of the interaction set as the ratio of the total energy and the distance between the particles by setting the observation point value of r_2 at the classical radius of the electron

$$r_2 = \alpha D_e \qquad (11)$$

The value of the potential energy of the interaction is defined it terms of the distance between the particles

$$\Delta\varepsilon = \frac{Q^2}{r_1} \qquad (12)$$

This symmetrizing of the interaction Eq.(9), defines the charge-charge elation between particles in the classical relationsip.

The potential energy is then

$$\Delta\varepsilon = \frac{Q^2}{r_1} \qquad (13)$$

Where r_1 is the distance between the classical radii of the paticles

$$\alpha = \frac{\sqrt{2}\,\lambda_{PL}}{\lambda_e}\,v_e \qquad (14)$$

or

$$\frac{\Delta\varepsilon}{\varepsilon_T} = \left(\frac{\lambda_{PL}^2}{\lambda_e^2}\sqrt{2}v_e\right)\left(\frac{\lambda_{PL}^2}{\lambda_e^2}\sqrt{2}v_e\right)\frac{\lambda_{e1}}{r_1}\frac{\lambda_{e2}}{r_2} \qquad (15)$$

For this to be correct the right side must be:

$$\frac{\Delta\varepsilon}{\varepsilon_T} = \frac{\alpha\lambdabar_{e1}}{r_1}\frac{\alpha\lambdabar_{e2}}{r_2} \qquad (16)$$

Thus in Eq.(15), the brackets have to be the fine structure constant α, thus:

$$\alpha = \left(\frac{\lambdabar_{PL}^2}{\lambdabar_e^2}\sqrt{2}v_e\right) = \frac{1}{137.0359997} \qquad (17)$$

(Detailed calculations in [4])

r_1 and r_2 Are the distances from the particle origin to the observation point, thus we can pick the observation point at the classical radius of the electron leaving the expression to be the potential of the second particle in the Feynman field of the second with the same minimum value

$$\frac{\alpha\lambdabar_{e1}}{r_1} = 1 \qquad (18)$$

and

$$\frac{\Delta\varepsilon}{\varepsilon_T} = \frac{1}{mc^2}\frac{Q^2}{r_2}* \qquad (19)$$

*This is the standard definition of potential energy for an electron:

*Note that the issue of the charge sign of the electric charges has not been addressed here, and is left to a subsequent paper. It could be conventionally assigned in Eq.(15), but it is more complicated than that, and has to do with the structure of the electron.

The coincidence between the calculations for both electric and gravitational interactions, and accuracy of the relation between G and α gives a degree of confidence that this procedure has merit.

From Eq.(8), and Eq.(17), The fine structure constant in terms of the Compton frequency, radius, and the gravitational radius of the electron is*:

$$\alpha = \sqrt{2}\,\frac{\mu_e \nu_e}{\lambda_e} \tag{20}$$

*See [4], for calculation details.

References:

Preceding reference Papers by Author

1. DT Froedge, The Connection between Electric Charge, Gravitation, and the Feynman Sum over All Histories View of Quantum Electrodynamics, April 2020 Conference: APS April. 18-21, 2020 Washington DC.
https://absuploads.aps.org/presentation.cfm?pid=18355 https://www.researchgate.net/publication/341310206

2. DT Froedge, A Quantum Theory Conjecture on the Origin of Gravitational and Electric Particle Interaction, December 2019, DOI: 10.13140/RG.2.2.29097.54884
https://www.researchgate.net/publication/337826826

3. DT Froedge, The Dirac Equation and the two Photon Model of the Electron
February 2021, https://www.researchgate.net/publication/349089256

4. DT Froedge, The Gravitational Constant to Eleven Significant Digits,
March 2020, DOI: 10.13140/RG.2.2.32159.38564
https://www.researchgate.net/publication/339943651

5. DT Froedge, The Dirac Equation and the two Photon Model of the Electron
https://www.researchgate.net/publication/349089256

6. Roger Blandford, Kip S. Thorne, Applications of Classical Physics, (in preparation, 2004), Chapter 26
http://pmaweb.caltech.edu/Courses/ph136/yr2012/1227.1.K.pdf

7. I. Shapiro; Gordon H. Pettengill; Michael E. Ash; Melvin L. Stone; et al. (1968). "Fourth Test of General Relativity: Preliminary Results". Physical Review Letters. 20 (22): 1265–1269. Bibcode:1968PhRvL..20.1265S. doi:10.1103/PhysRevLett.20.1265.

The Dirac Equation and the Two Photon Model of the Electron

D.T. Froedge

V041621

Abstract

In a previous paper, "The Electron as a Composition of Two Vacuum Polarization Confined Photons", included here as Appendix I [1]. The model shows structure, and the physical description required for the existence of an electron being a composition of two photons bound by the self-induced vacuum polarization gradient of the index of refraction. This physical model of the electron makes the mathematics more compatible with Lorentzian mechanics, and forms the basis for the interaction between the Feynman's sum-over-paths and electric charge.

This paper and its revision will focus on the mathematical structure of the electron including the spin aspects of its bound structure.

Background and Basis

Dirac

The Dirac expression factored by the use of Geometrical Algebra factors the wavefunction into a product of functions. This split is somewhat artificial in that the solutions are factors of a single wavefunction of two particles. The factored solutions have some unrealistic properties including negative energy and imaginary components that should be real. The problems primarily results from the roots of negative quantities. The equation is created out the Klein Gordon Equation with an additional requirement to satisfy the symmetry of spacetime, and there is no physical basis for the existence of the equation as a defining equation for the electron.

P-Dirac

The approach here is to first define wavefunctions of the individual photons in the electron having 4-derivitive that are proper Lorentz 4-vectors momentum. The magnitude of this sum can then be shown to have the properties of the electron including mass and spin.

The magnitude of the sum of two 4-vectors is the root of the square of the sum, thus if two proper Lorentz 4-vectors are summed the magnitude is the products are a Lorentz scalar invariant. For a single photon this is zero, but if there is a change in the index if refraction the magnitude is not zero, and there is rest mass component [2]. With the proper values of energy, and phases, the electron as a Lorentz scalar invariant rest energy as the sum of two photons can be derived.

This legitimately creates an equation for two bound photons from the well-known relativistic dynamics, having a proper rest mass and spin.

Photon Wavefunctions

The photon wavefunctions of the two photons are proposed to be a space-time function such that:

$$\psi_1 = e^{i\left(\frac{\hat{k}gx}{D_1} - \omega_1 t\right)} = e^{i\left(\frac{m_1 c_1}{h}\hat{k}gx - \frac{m_1 c_0}{h}ct\right)} \qquad \psi_2 = e^{i\left(-\frac{\hat{k}gx}{D_2} - \omega_2 t\right)} = e^{i\left(-\frac{m_2 c_1}{h}\hat{k}gx - \frac{m_2 c}{h}c_0 t\right)} \quad (1)$$

Where m is defined as $m = h\omega/c^2$ and $\hat{k}$ is the unit vector in the direction of travel. The photons are set, as shown, to be phase conjugates. The subscripts on c_0 refers to the velocity of light referenced to the observers frame and c_1, is the velocity of light in another frame or medium with an index of refraction different from the observer. [3],[4],[5],[6]. A change in the index of refraction c_1/c_0 does not change the frequency of the time component of the wavefunction.

The momentum 4-vector for the photon is the four derivative of the wavefunction:

$$p^\mu = h\gamma^\mu \frac{\partial}{\psi_1 \partial x^\mu} e^{i\left(\frac{mc_1}{h}\hat{k}gx - \frac{mc_0}{h}c_0 t\right)} \tag{2}$$

The four space vector p^μ is the standard four-space relativistic momentum In the Weil representation normally used to define the 4-space momentum of a photon. With, $c_1 = c_0$, the relativistic scalar magnitude is zero:

$$E_I = hc \left| \gamma^\mu \frac{\partial}{\psi \partial x^\mu} e^{i\left(\frac{mc_1}{h}\hat{k}gx - \frac{m_1 c_0}{h}ct\right)} \right| = 0 \tag{3}$$

The magnitude is the square root of the product of the vector times its conjugate.

If $c_1 = c_0$ the value of E_I vanishes thus there is no invariant rest energy, but if the location of the photon moves into a higher index of refraction such as piece of glass where $c_1 \neq c_0$ then the product or square of the four vectors is not zero, but a Lorentz invariant scalar [2], and the photon has not only relativist kinetic energy, but also a scalar invariant rest energy. This invariant rest energy creates rest mass equivalent in the rest frame of the glass. When it leaves the glass, the energy is again, pure relativistic energy, thus the glass has by slowing the velocity temporarily changed the energy from relativistic to invariant.

Though the wavefunctions may be strange to some, Eq.(2), and Eq.(3), are well understood Quantum methods for evaluation the momentum and energy of a photon.

The value of the invariant rest energy of a photon can be shown from the scalier magnitude of the resulting 4-momentum. Thus if the photon is in a location with a reduced value of c:

$$\Delta c_1 = c_0 - c_1 \quad \rightarrow \quad c_1 = c_0 - \Delta c_1 \tag{4}$$

and

$$\overset{r}{p_1} = h \left[\gamma^\mu \frac{\partial}{\psi \partial x^\mu} e^{i\left(\frac{m_1 c_0}{h}(x - c_0 t) - \frac{m_1 \Delta c_1}{h} x\right)} \right] \tag{5}$$

$$= \left| h \left(\gamma^\mu \left(\frac{m_1 c_0}{h} - \frac{m_1 c_0}{h} \right) - \gamma^k \frac{m_1 \Delta c_1}{h} \right) \right| = hc \left(\frac{m_1 \Delta c_1}{h} \right) \tag{6}$$

The product of two Lorentz four-vectors is a Lorentz scalar, [2], thus the magnitude of the invariant mass energy in Eq.(6), is added to the glass in its rest frame is:

$$E_I = m_1 c_0 \Delta c_1 \tag{7}$$

This is an important concept since it is clear that if a photon is slowed down or trapped in a volume of space, there is an invariant scalar rest mass present.

Three Parts of the Energy of a Photon

From the above discussion, the components of the total energy are; the invariant or rest energy E_I, the kinetic energy E_K, and the total energy, E_T, specifically this is:

$$E_I = c_0 h |p^\mu| = c_0 h \left| \gamma^\mu \frac{\partial \psi}{\psi \partial x} \right| \qquad E_K = c_0 h \left| \gamma^k \frac{\partial \psi}{\psi \partial x} \right| \qquad E_T = c_0 h \left| \gamma^0 \frac{\partial \psi}{\psi \partial x} \right| \tag{8}$$

The difference of the kinetic and total energy is the Lorentz invariant scalar rest mass.

Summing Two Photon Momentum

The sum of two Lorentz 4-vectors is also a Lorentz 4-vector, thus the magnitude of the momentum of two photons is a scalar invariant.

Presume $p_1 + p_2$ is the sum of the momentum for two photons, then the invariant energy for the sum is:

$$E_I = c_0 h \left| (p + p) \right| \tag{9}$$

or from Eq. (8):

$$E = c_0 h \left| \gamma^\mu \frac{\partial \psi_1}{\psi_1 \partial x} + \gamma^\mu \frac{\partial \psi_2}{\psi_1 \partial x} \right| \tag{10}$$

Consider the two identical phase conjugate photons of Eq. (1), which are moving in opposite directions:

$$p_1 = h \left[\gamma^\mu \frac{\partial}{\partial x^\mu} e^{i \left(\frac{m_1 c_0}{h} (kgx - c_0 t) \right)} \right] \qquad p_2 = h \left[\gamma^\mu \frac{\partial}{\partial x^\mu} e^{i \left(\frac{m_2 c_0}{h} (-kgx - c_0 t) \right)} \right] \tag{11}$$

The invariant energy of the sum is:

$$E_I = ch \left| (p_1 + p_2) \right| \tag{12}$$

or

$$E_I = ch \sqrt{(p_1 + p_2)(p_1 + p_2)^*} = ch \sqrt{\left(p_1 p_1^* + p_2 p_2^* + p_2 p_1^* + p_1 p_2^* \right)} \tag{13}$$

Both are Lorentz 4-vectors with a zero scalar value thus.

$$p_1 p_1^* = 0 \qquad\qquad p_2 p_2^* = 0 \tag{14}$$

The cross terms however are not necessarily zero, but are Lorentz scalar invariants [2]. If $p_1^k g p_2^{k*}$ is positive or the photons are in opposite directions, there is a scalar invariant rest mass. $(\gamma^k g \gamma^k = -1)$

$$p_1 p_2^* = \left(p_1^k g p_2^{k*} + p_1^0 p_2^{0*}\right) = 2m_1 m_2 c_0^2$$
$$p_2 p_1^* = \left(p_2^k g p_1^{k*} + p_2^0 p_1^{0*}\right) = 2m_1 m_2 c_0^2 \tag{15}$$

In this case the cross terms are alike and:

$$p_1 p_2^* = p_2 p_1^* \tag{16}$$

If the value of m_1 and m_2 is equal to the two internal photons in the electron [1] $m_e = m_1 + m_2$ then:

$$p_1 p_2^* + p_2 p_1^* = m_e^2 c_0^2 \tag{17}$$

The energy is then:

$$E_I = ch\sqrt{p_1 p_1^* + p_2 p_2^* + p_1 p_2^* + p_2 p_1^*} = \sqrt{4m_1 m_2}c_0^2 = m_e c_0^2 \tag{18}$$

Thus:

$$E_I = ch\sqrt{2p_1 p_2^*} = m_e c_0^2 \tag{19}$$

The value of the invariant scalar rest mass of the sum of the two opposing conjugate phase photons is that of the electron. If the two photons are confined such as in bound orbital motion as in the electron (Noted appendix I), this is a Lorentz invariant scalar measurable rest mass.

Connection to the Dirac Equation

Eq. (19), and Eq. (11), could now be restated as:

$$2\gamma^\mu \frac{\partial \psi_1}{\psi_1 \partial x} \times \gamma^\mu \frac{\partial \psi_2^*}{\psi_2^* \partial x} = \frac{m_e^2 c_0^2}{h^2} \tag{20}$$

This is the magnitude of the sum of the momentum of two conjugate phase photons, having energy equivalent to the electron, and is identical to the factored Dirac expression.

Dirac Equation

$$\left(\gamma^\mu \frac{\partial}{\partial x^\mu} + \frac{m_e c}{h}\right)\psi_1 \times \left(\gamma^\mu \frac{\partial}{\partial x^\mu} - \frac{m_e c}{h}\right)\psi_2 = 0$$

$$\text{or} \quad \gamma^\mu \frac{\partial \psi_1}{\psi_1 \partial x^\mu} \times \gamma^\mu \frac{\partial \psi_2}{\psi_2 \partial x^\mu} = \frac{m_e^2 c_0^2}{h} \tag{21}$$

The Dirac equation Eq. (21), represents two mass particles one having a positive inertial mass, and one of which has a negative inertial mass. The negative mass thought by Dirac to be the positron, but the positron also has a positive inertial mass. [7],[8].

P-Dirac

Eq. (20), designated as P-Dirac is an alternate to the Dirac equation and is the basis of our earlier assertion that the electron is the binding of two rotating, opposite going phase conjugate photons. (See Included Paper). The Dirac equation, as well as the Schrodinger equation are cast as a starting point for the electron, and electron interaction, and do not have a physical basis. The P-Dirac equation is the magnitude of the sum of the 4-momentum of **two proper Lorentz relativistic photon momentums that have an invariant rest mass** [2].

Photon-Photon Interaction and Spin

As has been noted the photons in the electron are held in orbit by the gradient in the speed of light induced by the probable action paths of the other photon [1]. These mutually induced increases in the index of refraction bind the photons together and function as a potential between the photons, and creates the spin components of the eigen-functions.

The momentum of the photon, $p = mc$ is a function of c, thus the angular momentum of the photon is a function of the index of refraction. (See Appendix I)

This change in c can be included in the wave equation of the photon, (Eq.(6)), and thus is included in the momentum. The change in c is a space vector and kgx can accommodate this change

$$\gamma^k \frac{\partial}{\partial x^k} kgx \to p + \Delta p \qquad (22)$$

By noting that $c = c_0 - (c_0 - c) = c_0 - \Delta c$, the change can be included in the momentum of the photons as noted in Eq.(6). If this is included in both p_1 and p_2. The properties of the internal properties of the electron can be developed as:

$$p_1 p_2{}^* = -\frac{m_1 c_0}{h} \left(\left(\gamma^k - \gamma^0 \right) - \gamma_{1\perp}^k \frac{\Delta c}{c} \right) \frac{m_2 c}{h} \left(\left(+\gamma^k + \gamma^0 \right) - \gamma_{2\perp}^k \frac{\Delta c}{c} \right) \qquad (23)$$

$$p_2 p_1{}^* \to -\frac{m_2 c_0}{h} \left(\left(-\gamma^k - \gamma^0 \right) + \gamma_{2\perp}^k \frac{\Delta c}{c} \right) \frac{m_1 c_0}{h} \left(\left(-\gamma^k + \gamma^0 \right) + \gamma_{1\perp}^k \frac{\Delta c}{c} \right) \qquad (24)$$

For the change in the momentum of one photon by the other as they turn in a circle the change c is perpendicular to p Thus we can designate $\gamma_{1\perp}^k$ being perpendicular to γ^k.

As shown earlier the change in c is proportional to the electric potential thus the relation of the binding force to an electric equivalent can be evaluated and shown equal to the Schwinger electric energy density.

Detailed of the calculations are included in Appendix I

$$p_1 p_2{}^* = \frac{1}{2} m_e^2 c_0^2 \left(1 + \frac{\Delta c}{c_0} \boldsymbol{\sigma} \right) \qquad (25)$$

$$p_2 p_1{}^* = \frac{1}{2} m_e^2 c_0^2 \left(1 - \frac{\Delta c}{c} \boldsymbol{\sigma} \right) \qquad (26)$$

The sum is the same as Eq. (17), Which is the expected value of the rest mass,

$$p_2 p_1^* + p_2 p_1^* = m_e^2 c_0^2 \tag{27}$$

The commutator of the interaction is a time independent spin vector

$$p_2 p_1^* - p_2 p_1^* = m_e^2 c_0^2 \frac{\Delta c}{c_0} \boldsymbol{\sigma} \tag{28}$$

The interaction terms of the magnitude of the momentum, Eq. (25), and Eq. (26), are just the two component eigenfunctions of the Dirac equation with the change in c, Δc, induced by the vacuum polarization.

Noting that from the Attached Paper, regarding the two photon model, Eq. (1,17), the change in c at the electron orbit as a result of the spin and index of refraction to bind the photon is:

$$\frac{\Delta c}{c_0} = \frac{1}{2} \tag{29}$$

The commutator of the two eigenfunctions is then:

$$p_2 p_1^* - p_2 p_1^* = \frac{1}{2} m_e^2 c_0^2 \sigma \tag{30}$$

Spin in the Dirac and the P-Dirac Equations

Dirac was able to deduce the spin and angular momentum of the electron by applying an electromagnetic field to the equation and noting the difference in the effect on the two eigenfunctions, as a result of a potential generated by the angular momentum. This separated the effect of the angular momentum from the total momentum allowing evaluation.

In the two photon model it is easy to see the contribution of the angular momentum to each of the photon interaction terms Eq. (25), and Eq. (26). Additionally, from the two photon model, the spin can calculated directly (Attached Paper [1]), Eq. (1,8))

Discussion

The eigenfunctions of the two primary solutions of the P-Dirac equation $p_2 p_1^*$ and $p_1 p_2^*$ are directly identified with of the Dirac eigenfunctions solutions each of which have the same spin as found by Dirac. The mass energy for both the photon interaction terms is positive, whereas for the Dirac expression one is negative.

The relation between the Dirac equation Eq.(21), and the two photon Equation developed here is clear. The Dirac equation is an operation on a single wavefunction for two particles; it has two solutions one of which has negative energy, and an imaginary electric moment. It is suggested here that the problem is that the single function solution requires the presence of square roots which generate negative and imaginary values in the solutions.

The solution presented here, (P-Dirac) is of the product of two photon momenta constituted as the 4 gradient of well-defined photon wavefunctions. By reducing the velocity (or increasing the index of refraction) of the photons by a change in c necessary to provide a stable orbit generated by the Schwinger vacuum polarization, an additional spin term occurs in each of the interaction eigenfunctions. These terms are identical to those in each the Dirac solutions. The terms are opposite thus for the sum the interaction terms the spin vanishes, but the commutator is the sum of these two spin matrix indicating it is silent unless the particle is subjected to the presence of an electromagnetic field. Including the electromagnetic four-potential the photon momentum of Eq.(23), or Eq.(24), produces the same fictitious moments.

References:

1. DT Froedge, The Electron as a Composition of Two Vacuum Polarization Confined Photons Revised Apr. 2021 DOI: 10.13140/RG.2.2.18971.18722 https://www.researchgate.net/publication/350740864

1a. DT Froedge, A Physical Electron-Positron Model in Geometric Algebra, V041917 http://www.arxdtf.org/css/electron.pdf, http://vixra.org/pdf/1703.0274v2.pdf

2. P. Avery, Relativistic Kinematics II PHZ4390, Aug. 26, 2015, http://www.phys.ufl.edu/~avery/course/4390/f2015/lectures/relativistic_kinematics_2.pdf

3. B. Smith, Photon wave functions, wave-packet quantization of light, and coherence theory, http://iopscience.iop.org/article/10.1088/1367-2630/9/11/414

4. I, Bialynicki-Birula Photon Wave Function, Progress In Optics XXXVI, pp. 245-294 arXiv:quant-ph/0508202v1 26 Aug 2005 https://arxiv.org/abs/quant-ph/0508202v1

5. B. Smith, Photon wave functions, wave-packet quantization of light, and coherence theory, http://iopscience.iop.org/article/10.1088/1367-2630/9/11

6. I, Bialynicki-Birulaphoton Photon Wave Function, Progress In Optics XXXVI, pp. 245-294 arXiv:quantph/0508202v1 26 Aug 2005

7. W. Heisenberg and H. Euler, "Folgerungen aus der Diracschen Theorie des Positrons", Zeitschrift für Physik, 98 (1936) pp. 714-732. doi:10.1007/BF01343663 (https://dx.doi.org/10.1007%2FBF01343663) English translation (http://arxiv.org/abs/physics/0605038)

8. L. Rosenfeld Theory of Electrons (North Holland Publishing Company, Amsterdam) 1951

Appendix I

Detail Calculations of the Photon Cross Terms of $p_1p_2{}^*$ and $p_2p_1{}^*$

$$p_1p_2{}^*$$

$$p_1p_2{}^* = -\frac{m_1c_0}{h}\frac{m_2c}{h}\left(\left(\gamma^k - \gamma^0\right) - \gamma_{1\perp}^k\frac{\Delta c}{c}\right)\left(\left(+\gamma^k + \gamma^0\right) - \gamma_{2\perp}^k\frac{\Delta c}{c}\right) \quad (31)$$

$$p_1p_2{}^* = -h^2\frac{m_1c_0}{h}\frac{m_2c}{h}\left[\begin{array}{c} +\left(\gamma^k - \gamma^0\right)\left(+\gamma^k + \gamma^0\right) + \left(\gamma_{1\perp}^k\frac{\Delta c}{c}\gamma_{2\perp}^k\frac{\Delta c}{c}\right) \\ -\gamma_{1\perp}^k\frac{\Delta c}{c}\left(+\gamma^k + \gamma^0\right) - \left(\gamma^k - \gamma^0\right)\gamma_{2\perp}^k\frac{\Delta c}{c} \end{array}\right]$$

$$p_1p_2{}^* = -\frac{1}{4}m_e^2c_0^2\left[\begin{array}{c} -2 + \frac{\Delta c}{c}\frac{\Delta c}{c}\gamma_{1\perp}^k\gamma_{2\perp}^k \\ +\frac{\Delta c}{c}\left[-\left(+\gamma_{1\perp}^k\gamma^k + \gamma_{1\perp}^k\gamma^0\right) - \left(\gamma^k\gamma_{2\perp}^k - \gamma^0\gamma_{2\perp}^k\right)\right] \end{array}\right] \quad (32)$$

Note that in the circular orbit of eh photons $+\gamma_{1\perp}^k$ is perpendicular to γ^k, the photon momentum thus, $\gamma_{1\perp}^k\gamma^k = 0$, and the space vector and the time vector anti-commute thus, $\gamma_{1\perp}^k\gamma^0 = -\gamma_{2\perp}^k\gamma^0$. The product of $\gamma_{2\perp}^k\gamma^0 = \boldsymbol{\sigma}$, is the Dirac spin vector.

The value of the square is: $\left(\frac{\Delta c}{c}\frac{\Delta c}{c}\right) = 2\frac{\Delta c}{c}$, thus the Expression becomes:

$$p_1p_2{}^* = \frac{1}{2}m_e^2c_0^2\left(1 + \frac{\Delta c}{c_0}\boldsymbol{\sigma}\right) \quad (33)$$

$$p_2p_1{}^*$$

With the same procedures for $p_2p_1{}^*$ the results are:

$$p_2p_1{}^* \rightarrow -\frac{m_2c_0}{h}\frac{m_1c_0}{h}\left(\left(-\gamma^k - \gamma^0\right) + \gamma_{2\perp}^k\frac{\Delta c}{c}\right)\left(\left(-\gamma^k + \gamma^0\right) + \gamma_{1\perp}^k\frac{\Delta c}{c}\right) \quad (34)$$

or:

$$p_2 p_1^* \rightarrow -\frac{1}{4} m_e^2 c_0^2 \left(-\left(2 + 2\frac{\Delta c}{c}\right) + 2\frac{\Delta c}{c} \gamma_{2\perp}^k \gamma^0 \right) \qquad (35)$$

and

$$p_2 p_1^* \rightarrow -\frac{1}{4} m_e^2 c_0^2 \left(-\left(2 + 2\frac{\Delta c}{c}\right) + 2\frac{\Delta c}{c} \gamma_{2\perp}^k \gamma^0 \right) \qquad (36)$$

$$p_2 p_1^* = \frac{1}{2} m_e^2 c_0^2 \left(1 - \frac{\Delta c}{c} \boldsymbol{\sigma} \right) \qquad (37)$$

Attached Paper

Included Below

Attached Paper

The Electron as a Composition of Two Vacuum Polarization Confined Photons Revised II

This is the second revision of this paper and it resulting from discovering some errors when developing the two photon version of the Dirac equation (P-Dirac)

D.T. Froedge
V022620

Abstract

In a previous paper "A Physical Electron-Positron Model"[1] an electron model was developed in a geometrical algebra (GA) construct developed by Doran et.al. [2] The model shows the mathematical structure, and the physical description required for the existence of an electron as a composition of two photons bound by the self-induced vacuum polarization gradient of the index of refraction. This paper will develop the mechanics of a vacuum polarization induced index of refraction, binding photons in orbit around a common center of momentum.

The concept of charge has heretofore not had a theoretical explanation, accept for some unknown substance associated with mass. This model offers the physical concept of charge created from QFT mechanics.

If a plane polarized photon wavefunction is turned into a circle by a gradient in the index of refraction induced by a second photon, and this circle circumference is half the wavelength, then the wave polarization is constant along the radial axis, and the photon electric vector can be said to be radially polarized.

Introduction

The wave particle duality of particle dynamics is understood as physical aspects of particles that require both perspectives to predict the outcome of experimental tests. For the purposes of this paper we will subscribe to the wave nature of a photon as a prediction of the probability location, and the particle as a point particle with dynamics directed by a gradient in the speed of light induced by the nonlinear aspects of vacuum polarization. The physical photon is assumed to be very small, on the order of the Planck particle.

The wave nature of the electron has been well developed by Schrodinger, Dirac and many others. The Lagrangian wave nature alone however is inadequate to describe well known measurable phenomena such as charge, size and mass

By appealing to the particle nature and the nonlinear effects of photon interaction, a composition particle that has, mass, spin, and size can be developed. The electron size has been a particularly difficult issue for QFT since there is an infinite singularity associated with the electron. This model should be useful in regard to resolving some of those issues. The individual photons in the model still have singular aspects, but not the infinities associated with the electron.

In Geometric Algebra (GA) the Dirac Matrices becomes the spacetime unit coordinate vectors, which indirectly changes the normal view of QM by defining some of the aspects QM as actually features of Lorentz covariant spacetime. Parity, time reversal, and mass, become part of the spacetime structure, simplifying the mapping of the Dirac relativistic quantum representation into the eight dimensional, subalgebra of the GA spacetime representation. This allows a GA functional description of a photon. [1], which in turn allows a four-dimensional composite electron.

The authors previous paper [1], proposed a model of an electron formulated as the composition of two photons using the AG rotor structures for QM formulated by Doran et.al. [2]. (Fig. 1).

This presentation offers an electron model in that context having similarities of the atomic physical model, but relies on the gradients in the index of refraction produced by the nonlinear effects of vacuum polarization as the binding mechanism.

Sketches

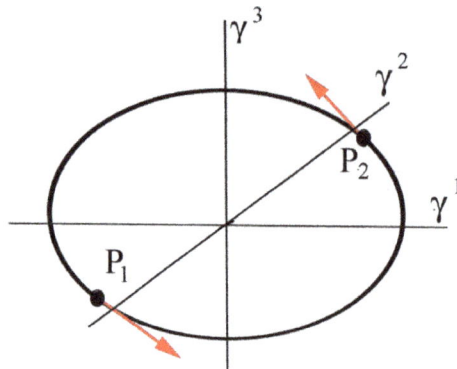

Fig.1 General configuration showing the orbiting of photons in a GA cooridnate syatem

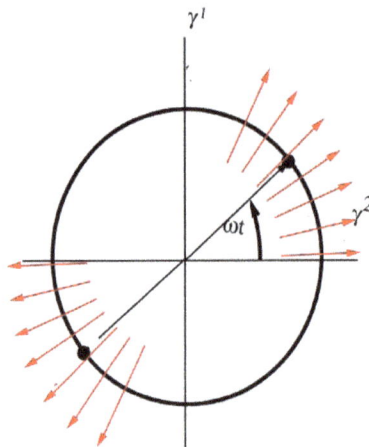

Fig.2. Radially polarized photons bound by the self-generated vacuum polarization gradient, maintaining a radially polarized circular electric vector probability.

The electric vector does not constitute a charge, but a polarization of the photon flow probability. The Lorentz transformation properties that constitute the magnetic vectors are determined by its internal rotations and are opposite for photons of opposite chirality.

Structure of the article

Primary Physical Mechanisms

The interaction of the two orbiting photons (Fig.1) as described in the earlier paper is not the result of an electric force, but by the motion under the influence of the mutual gradient in the index of refraction. This gradient is generated by the self-induced vacuum polarization of the two opposite going photons.

The vacuum polarization effect between two interaction photons is a well-researched process both from theoretical and experimental aspects. The first development by Sauter, Serber, Euler and others,[3],[4],[5], and later by more sophisticated methods of QFT by Schwinger and others[8],[9].

The study of the vacuum polarization on the index of refraction is quite extensive in the lower levels of E when birefringence on photons in static fields effects are predominant, [10],[11],[12],[13] ,[14],[15],[16],[17],[18], [19],[20],[21]. Others have studied and proposed experiments investigated the effects of photon-photon scattering in the higher energy levels, ,[22],[23], and there have been several proposals for studies on the effects on the index of refraction by intense laser beams,[24],[25],[26],[27],[28],[29].

At the higher E levels, that are more appropriate to this work, the processes of Delbruck scattering and pair production dominates. These processes, originally proposed by Max Delbruck and first observed, by Robert Wilson [30], have been the subject of intensive research in both theoretical, and experimental since the 1950's. [31],[32], [33],[34],[35],[36], [37]. Appropriate to this paper, but not at the same energy levels is the research done by J. Kim et.al. [38], on light bending in a Coulombic field.

Vacuum polarization and the Index of Refraction

The most important aspect has been the derivation by Schwinger of the leading nonlinear corrections to the vacuum polarization that allows calculations of the local index of refraction below the critical electron–positron limit [3].

$$E_{CR} = \frac{m_e^2 c^3}{Q\hbar} = \frac{c\hbar}{Q \lambda_e^2} \qquad (1,1)$$

At the low-energy end with non-parallel fields generally defined by the Heisenberg-Euler Lagrangian are the studies of birefringence changes in the index of refraction induced at low levels $(E \ll E_{cr})$ .These have been conducted by a large number of researchers [10-21], and the results are generically similar to:

$$\eta^{P,\perp} = 1 + \frac{\alpha(11\,m\,3)}{45\pi} \frac{E_2^2}{E_{cr}^2} \qquad (1,2)$$

The $P,\perp$ suffix indicates parallel and perpendicular field polarizations.

For two photons moving around a common center of momentum each experiences the electromagnetic field of the other. The relation for that interaction at $E \ll E_{CR}$ from Kim et.al, "Light bending in radiation background" [38], and Light bending in a Coulombic field the index of refraction can be expressed as:

$$\eta^{-1} = \frac{c}{c_0} = 1 - \frac{(14 \perp, 8P)\alpha^2 h^3}{45 m^4 c^5}(u \times E_2)^2 \tag{1,3}$$

At the higher end of the energy levels above the Kim et.al, work closer to the Schwinger limit $(E : E_{cr})$, the index of refraction is better understood and by the processes related to Delbruck scattering, and pair production.

The reflection coefficient expressed in the relative index of refraction and the high end scattering experiments, lead to the conclusion that the index of refraction has an infinity at the Schwinger Limit. With multiple loops and higher order corrections the index of refraction at the higher fields as developed by Dietrich et.al. ([12] Fig,1) is: $E \rightarrow E_{cr}$ index of refraction η^{-1} is:

$$\sqrt{\eta^{-2}} \rightarrow \left(1 - Q\frac{B^2}{2B_{cr}^2}\sin^2\theta\right) \tag{1,4}$$

At very high levels of the fields the Q factor $\rightarrow 1/2$, and if E represent the maximum of the E & B fields of photons near the Schwinger limit then for two opposite colliding photons with fields of E_1 and E_2 the maximum local index of refraction is:

$$\eta^{-1} \rightarrow \left(1 - \frac{E_1 \cdot E_2}{2E_{cr}^2}\right) \tag{1,5}$$

Index of Refraction for Photons containment

The index of refraction to maintain photons in a circular path can be determined from classical physics by variation methods applied to Fermat's principle. It is straight forward and well done by J. Evans, et.al. [39], and for stable orbits Fermat's principle requites the index of refraction to be proportional to$1/r$, thus in terms of the Compton radius for an electron, This can be written as:

$$\frac{c}{c_0} = \frac{kr}{\lambda_e} \quad \rightarrow \quad \frac{\Delta c}{c_0} = 1 - \frac{kr}{\lambda_e} \tag{1,6}$$

k is the index of refraction constant and λ_e is the Compton radius of the electron

Putting this into Eq.$(1,5)$, the relations between c, r and E is:

$$\eta^{-1} = \left(\frac{c}{c_0}\right) = \left(\frac{kr}{\lambda_e}\right) = \left(1 - \frac{E^2}{E_{cr}^2}\right) \tag{1,7}$$

The value of k can be determined from Eq.$(1,6)$, and by appealing to the Electron spin of the two orbiting photons that make up the electron.

$$(m_1 + m_2)cr = \frac{\hbar}{2} \rightarrow cr = \frac{\hbar}{2m_e} \rightarrow \frac{\hbar c_0}{2m_e c_0} = \frac{\lambda_e c_0}{2} \tag{1,8}$$

$$\tag{1,9}$$

The momentum $p = mc$, as well as the angular momentum are a function of c thus the index of refraction η^{-1} is

$$\frac{c}{c_0} = \frac{\lambda_e}{2r} \tag{1,10}$$

The ratio of c/c_0 from the requirement for stable photon orbits Eq.$(1,6)$, and the electron spin and Eq.$(1,10)$, must be equal at the photon primary orbit.

$$\frac{c}{c_0} = \frac{\lambda_e}{2r} \quad , \quad \frac{r}{2\lambda_e} = \frac{c}{c_0} \qquad (1,11)$$

Eq.(1,6), and Eq.(1,10), have a solution at $k = 1/2$, and the primary orbit for the photons r_p is the Compton radius for the electron.

$$r_p = \lambda_e \qquad (1,12)$$

From Eq.(1,6), At the primary orbit the value of Δc is :

$$\frac{\Delta c}{c_0} = \frac{1}{2} \qquad (1,13)$$

The two photons orbit at the Compton radius of the electron, exactly the expected radius of the electron, (Fig. 3.).

Photon Velocity vs Orbital Radius

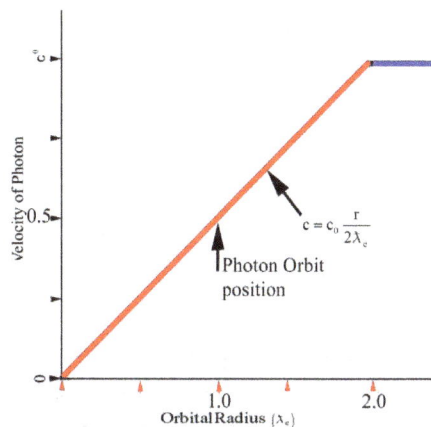

Fig. 3. This is the velocity of light experienced by the photons in orbit around the center of momentum as the result of vacuum polarization, and the most probable position of the orbit.

Connection to Schwinger Vacuum Polarization

If the centrifugal force on the orbiting photons is set equal to an electric force, the electric field in the electric field intensity in the electron can be evaluated from the centrifugal force and the orbital radius. Eq.(1,12), the radius is the Compton radius of the electron, thus.

$$f_c = \frac{m_1 c^2}{r} \rightarrow \frac{mc^2}{\dfrac{h}{m_e c_0}} =\rightarrow \frac{m_2 c^3}{h} \tag{1,14}$$

If this is then set to be the electric field force:

$$QE \rightarrow \frac{m_c^2 c^3}{h} \rightarrow E = \frac{m_c^2 c^3}{Qh} = E_{cr} \tag{1,15}$$

I The value of E ,or the electric density to hold the photon in place if it were charged is the **same as the value of the Schwinger critical Electric intensity for pair production** Eq.(1,1). Putting this into Eq.(1,5), gives the value of the index of refraction which is the index of refraction at the photon orbit:

$$\eta^{-1} \rightarrow \left(1 - \frac{E_{cr}^2}{2E_{cr}^2}\right) = \frac{1}{2} \tag{1,16}$$

For clarification of the relation between c and η^{-1}

$$\left(1 - \eta^{-1}\right) = \frac{c_0 - c}{c_0} = \frac{\Delta c}{c_0} = \frac{1}{2} \tag{1,17}$$

This calculation should not be taken too seriously since the change in c in the frame of the photon is actually due to the apparent increase in density of the background Feynman photons due to the photon orbital rotation.

Photon Radial Alignment: Thomas Precession

Thomas Precession is a well understood phenomenon totally within the mechanics of Lorentz dynamics.

As a particle rotates around a center axis there is a frame rotation such that when it arrives back at a defined point its helictical phase orientation will also have been rotated. For a photon in a circular index of refraction this will mean that for every cycle of rotation about the axis its helictical rotation will be reduced by that number of cycles. At half the Compton radius of the photon, this reduces the helictical frequency to zero leaving the photon with a constant radial electric vector. See **Appendix I** for details.)

Conclusion

A model for an electron has been presented that physically demonstrates mass, spin, & polarization within the concepts of currently known physics. Nothing has been postulated that isn't well understood in terms of current physics. Neither of the physical regimes of QM or Classical physics have been stretched, compromised, or extended beyond that which has been experimentally confirmed.

It gives a physical insight to the mechanical process, and since there are no singularities associated with the model. This structure allows QFT a path around the infamous renormalization without having to cancel infinite values with infinite values.

The Electron-Electron interactions as the result of the external Feynman Photon-photon interactions are included in "A Quantum Theory Conjecture on the Origin of Gravitational and Electric Particle Interaction"

References

1. DT Froedge, A Physical Electron-Positron Model in Geometric Algebra, V041917 http://www.arxdtf.org/css/electron.pdf, http://vixra.org/pdf/1703.0274v2.pdf

2. C. Doran, A. Lasenby 2003 Geometric Algebra for Physicists, Cambridge University Press, Quantum Theory and Spinors, chapter 8

3. Sauter, "Über das Verhalten eines Elektrons im homogenen elektrischen Feld nach der
relativistischen Theorie Diracs", Zeitschrift für Physik, 82 (1931) pp. 742–764.
doi:10.1007/BF01339461 (https://dx.doi.org/10.1007%2FBF01339461)

4. W. Heisenberg and H. Euler, "Folgerungen aus der Diracschen Theorie des Positrons",
Zeitschrift für Physik, 98 (1936) pp. 714-732. doi:10.1007/BF01343663 (https://dx.doi.org/10.1007%2FBF01343663) English translation (http://arxiv.org/abs/physics/0605038)

5.R. Serber, "Linear modifications in the Maxwell field equations," Phys. Rev. 48, 49 (1935); E. A. Uehling, "Polarization effects in the positron theory," Phys. Rev. 48, 55-63 (1935).

6. H. Euler and B. Kockel, "The scattering of light by light in the Dirac theory", Naturwiss. 23, 246 (1935).

7. H. Euler, "On the scattering of light by light in Dirac's theory", PhD thesis at Univ. Leipzig (1936); published in Ann.Phys. (Leipzig) 26, 398 (1936).

8. H. Euler, "On the scattering of light by light in Dirac's theory", PhD thesis at Univ. Leipzig (1936); published in Ann.Phys. (Leipzig) 26, 398 (1936).

9. W. Heisenberg and H. Euler, "Consequences of Dirac's Theory of Positrons", Zeit. f. Phys. 98, 714 (1936); an English translation is at arXiv:physics/0605038.

10. J Schwinger, On Gauge Invariance and vacuum polarization , Physical Review volume e82, 1951

11. L. Rosenfeld Theory of Electrons (North Holland Publishing Company,
Amsterdam) 1951 arXiv:0906.3018v1 [physics.gen-ph]

12. W Dittrich, H. Gies, Vacuum Birefringence in Strong MagneticFields, arXiv:hep-ph/9806417v1 19 Jun 1998

13. T. Heinz M. Platz, Observation of Vacuum Birefringence: A Proposal
https://www.tpi.uni-ena.de/qfphysics/homepage/wipf/publications/papers/nonlineared.pdf

4. A. Rikken and C. Rizzo, Magnetoelectric birefringences of the quantum vacuum, Phys. Rev. A, Vol.63, 012107 (2000)

15. J. Heyl, L. Hernquist, Birefringence and Dichroism of the QED Vacuum
Lick Observatory, 12.20.Ds, 42.25.Lc 97.60.Jd, 98.70.Rz

16. T. Heinz, Observation of Vacuum Birefringence: A Proposal, https://www.tpi.uni-jena.de/qfphysics/homepage/wipf/publications/papers/nonlineared.pdf, 2005

17. W. Tsai, T. Erber, Propagation of photons in homogeneous magnetic fields. Index of refraction, Physical Review, D Volume 12, Number 4 1975.

18. C. Turtur, A Hypothesis for the Speed of Propagation of Light in electric and magnetic fields and the Planning of an Experiment for its Verification. Wolfenbüttel, November - 23 - 2007

19. E.Seegert, Quantum Reflection at Strong Magnetic Fields, Masterarbeit zur Erlangung des akademischen Grades Master of Science (M.Sc.) geboren am 23.06.1986 in Berlin Matrikelnummer: Physikalisch-Astronomische Fakultät 2013

20. E. Zavattini, et. al., Experimental Observation of Optical Rotation Generated in Vacuum by a Magnetic Field, Phys. Rev. Lett 96, 110406 (2006)

21. M. Soljacic´ and M. Segev, Self-trapping of electromagnetic beams in vacuum supported by QED nonlinear effects Physical Review A, volume 62, 0438172000

22. R. Battesti, C. Rizzo, Magnetic and electric properties of quantum vacuum, Progress in Physics, IOP Publishing, 2013, https://hal.archives-ouvertes.fr/hal-00748532

23. A. E. Shabad, Interaction Of Electromagnetic Radiation With Supercritical Magnetic Field, International Workshop on Strong Magnetic Fields and Neutron Stars. https://arxiv.org/abs/hep-th/0307214

24. A. Akhieser, L. Landau and I. Pomeranchuk, "Scattering of light by light", Nature 138, 206 (1936); A. Akhieser, "'Uber

25. M. Born, L. Infeld, Die Streuung, Von Licht An Licht", PhD thesis, 1936, Ukrainian Physico-Technical Institute; Phys. Zeit. Sow. 11, 263 (1937).

26. J. Kim, T. Le; Light bending in radiation background: Journal of Cosmology and Astroparticle Physics; http://iopscience.iop.org/article/10.1088/1475- 516/2014/01/002/pdf

27. X. Sarazin1a, F. Couchot1, A. Djannati, Atai, O. Guilbaud3, S. Kazamias, M. Pittman, M. Urban, Refraction of light by light in vacuum, EPJ manuscript, arXiv:1507.07959v3 [physics.optics] 2015

28. M. Urban, F. Couchot, S. Dagoret-Campagne,X.R Sarazin Corpuscular description of the speed of light in a homogeneous medium16 Jun 2009 arXiv:0906.3018v1 [physics.gen-ph] 16 Jun 2009

29. M. Urban A particle mechanism for the index of refraction, LAL 07-79, 2007,
Dmitri Kharzeeva and Kirill Tuchinb,Vacuum self–focussing of very intense laser beams, BNL-NT-06/43, RBRC-657,arXiv:hep-ph/0611133v2 21 Feb 2007 https://arxiv.org/abs/0709.1550

30. M. Urban, F. Couchot, X. Sarazin, A. Djannati-Atai, The quantum vacuum as the origin of the speed of light, EPJ manuscript, arXiv:1302.6165v1 [physics.gen-ph] 2013

31. B. King1 and T. Heinzl, Measuring Vacuum Polarisation with High Power Lasers, Power Laser Science and Engineering, 2015, arXiv:1510.08456v1 [hep-ph] 28 Oct 2015

32. R. Wilson, Scattering of 1.33 Mev Gamma-Rays by an Electric Field
Physical Review, vol. 90, Issue 4, pp. 720-721, 1953

33. A. Scherdin, A. Schäfer, and W. Greiner, Delbrück scattering in a strong external field, physical review d volume 45, number 8 15, 1992

34. Delbruck scattering non-perturbative QED and possible applications
D. Habs Romania, Oct 03-05, 2012 1, Max-Planck-Institut f. Quantenoptik

35. T. Lee, Delbruck scattering cross section from light bending 2017, arXiv:1711.00160v1 [hep-ph]

36. W. Dittrich and H.Gies, Light propagation in non-trivial QED vacua arXiv:hep-ph/9804375v1 23 Apr 1998. https://arxiv.org/pdf/hep-ph/9804375.pdf

37. H.Gies1 G. Torgrimsson, Critical Schwinger pair production, 2015, arXiv:1507.07802v1 [hep-ph]

38. S. Akhmadaliev, et al., Experimental investigation of high-energy photon splitting in atomic fields, Phys. Rev. Lett. 89, 061802 (2002) [hepex/0111084].

39. Z. Bialynicka-Birula, I. Bialynicki-Birula, Nonlinear Effects in Quantum Electrodynamics. Photon Propagation and Photon Splitting, in an external Field. Phys. Rev. D, Vol.2, No.10, 1970

40. J. Kim, T. Le; Light bending in a Coulombic field: arXiv:1012.1134v2 [hep-ph] 7 Jan 201

41. J. Evans, M. Rosenquist, F=ma Optics, American Journal of Physics 54, 876 (1986); https://doi.org/10.1119/1.14861

43. S Weinberg, The Quantum Theory of Fields, Cambridge University Press,
p. 358, ISBN 0-521-55001-7, (1995)

43. P. A. M. Dirac, The Principles of Quantum Mechanics, 4th ed., Oxford University Press, London (1958).

44. Feynman, Hibbs, 1965, QuantumMechanics and Path Integrals McGraw-Hill

45. G. Dunne, The Heisenberg-Euler Effective Action,75years on, rXiv:1202.1557v1 [hep-th] 2012, arXiv:1202.1557v1 [hep-th] 7 Feb 2012

46. DT Froedge,A Quantum Theory Conjecture on the Origin of Gravitational and Electric Particle Interaction https://www.researchgate.net/publication/337826826_

Appendix I

Thomas Precession

As a pair of photons rotates around the center of a circle due to a variable index of refraction, the Thomas precession reduces the helictical rotation frequency of the photon. The photons frequency is reduced by exactly the axial frequency of the rotation. As the gradient in the index of refraction is increased the sum of the frequencies must remain constant.

As the circumference is reduced to the wavelength the helictical frequency is stopped. The rotation frequency is then equal to the original free particle frequency of the photon and the photon electromagnetic vectors are polarized along the orbital radius.

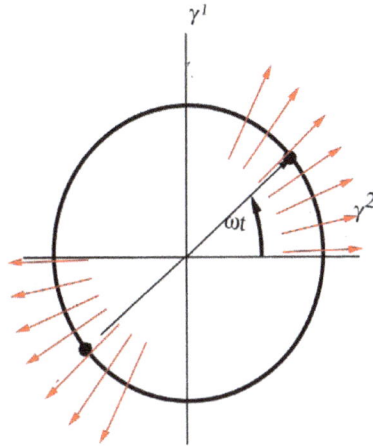

Fig.2 Two radially polarized photons bound by the self-generated vacuum polarization gradient in the index of refraction. The radially polarized circular directed electric vectors constitute the effect of a local charge.

This is easily shown from Lorentz geometric principles, the Thomas reduction to the frequency of an orbiting photon is:

$$\omega_T = \frac{1}{c^2}\left(\frac{\gamma^2}{\gamma+1}\right)\mathbf{a} \times \mathbf{v}$$

(1,18)

a is the circular acceleration dr / dt in the moving frame thus:

$$\Delta t' = \gamma \Delta t$$

(1,19)

and for a photon moving in a variable index of refraction the precession is:

$$\overset{r}{\omega}_T = \frac{1}{c^2}\left(\frac{\gamma^2}{\gamma+1}\right)\mathbf{a} \times \mathbf{v} \rightarrow \frac{1}{c^2}\left(\frac{\gamma^2}{\gamma+1}\right)\frac{d\mathbf{v}}{\gamma dt} \times \mathbf{v} = \frac{1}{c^2}\left(\frac{\gamma^2}{\gamma+1}\right)\frac{d\mathbf{v}}{dt} \times \mathbf{v}$$

For the photon the circular acceleration is:

$$\frac{dv}{dt} = \frac{c^2}{r}$$

(1,20)

121

And for the photon orbiting at the Compton radius:

$$r = \frac{c}{\omega_P} \qquad (1,21)$$

Thus as the radius is reduced to the Compton radius the Thomas precession frequency reduces the helictical frequency to zero, whereas the axial frequency in the orbit plane $\overset{r}{\omega_R}$ becomes equal to the free photon frequency.

$$\overset{r}{\omega_T} = \omega_P \uparrow \qquad (1,22)$$

The Thomas precession thus establishes a radial polarization if the electric vector along the radius vector to the center of momentum.

The Calculated Value of the Fine Structure Constant from Fundamental Constants

D.T. Froedge

V090921

Abstract

This paper is an extract from more extensive paper on Vacuum Polarization [1], from which the details of the calculation of the Fine Structure constant α may be obscured.

Main

The derived value of the constant from Eq.(31), and Eq.(32), of the paper is:

$$\alpha = \frac{\sqrt{2}\,\lambda_{PL}}{\lambda_e}\nu_e = \frac{\sqrt{G\hbar/2c}}{\pi(\lambda_e g_A)^2} \tag{1}$$

The fundamental constants in this expression are the Planck particle radius, the Compton radius of the electron, and the electron frequency ν_e. The frequency is the ratio in number of times a photon passes a point orbiting in the electron verses the number moving in a straight line. Numerically is equal to the Compton frequency of the electron. The Planck radius is $\lambda_{PL} = \sqrt{G\hbar/c^3}$. The g_A, constant is the precise Electron Magnetic Moment, $g_e / 2$, and the QED correction to the Compton radius of the electron.

The value of the fine structure constant, Alpha is:

$$\alpha = 1/137.035999710 \qquad (2)$$

The value of the gravitational constant is set to be $6.6755052700E\text{-}8$ (CGS). The Codata Gravitational Constant is a consensus value having about a tenth of a percent scatter. The value found here is within the error bars of all the measurements taken by the Bureau of International Weights and Measurements [19], and the BIPM* measurements are one of the measurement sets used in the consensus value.

The predicted value calculated here is thus testable to the experimental accuracy of the Gravitational constant

$$G = \frac{\alpha^2 2\pi^2 c \left(\lambda_e g_A\right)^4}{\hbar} = 6.6755052700E - 8 \qquad \text{(CGS)} \qquad (3)$$

Reference values of Constants and sources used in the calculation (CGS units)

$c = 2.9979245800E{+}10*$ $\qquad$ $g_A = g_e / 2 = 1.00115965218\dagger$

$\hbar = 1.054571817646E\text{-}27 *$ $\qquad$ $\lambda_e = \hbar / m_e c = 3.8615926794E\text{-}11$

$\alpha = 1/137.035999710(96) \dagger$ $\qquad$ $m_e = 9.1093837015(28)$

* Definition† Gabrielse et. al. [2][3][4]. ⊕ 2018 Codata Recommended Values

BIPM* and Codata consensus values (CGS)

Calculated value $\qquad$ G = 6.6755052700E-8
BIPM 01-32-2001 $\qquad$ G = 6.67559(27) x 10-8
BIPM weighted mean 2014 $\qquad$ G = 6.67554(16) x 10-8
BIPM Sep, 2015 $\qquad$ G = 6.67545(18)) x 10-8
Codata Consensus Value $\qquad$ G = 6.67430(15) x 10-8

*BIPM- Bureau of International Weights and Measurements, [5].

References:

1. DT Froedge, Vacuum Polarization, Gravitation, Charge, and the Speed of Light, Sep, 9, 2021, DOI:10.13140/RG.2.2.15619.22569, https://www.researchgate.net/publication/354474157

2. G. Gabrielse et al,, New Determination of the Fine Structure Constant from the Electron g Value and QED, Phys. Rev. Lett. 97, 030802 (2006)
http://hussle.harvard.edu/~gabrielse/gabrielse/papers/2006/NewFineStructureConstant.pdf

3 G. Gabrielse et al Cavity Control of a Single-Electron Quantum Cyclotron Measuring the Electron Magnetic Momen, arXiv:1009.4831v1 [physics.atom-ph] 24 Sep 2010

4. G. Gabrielse, New Measurement of the Electron Magnetic Moment and the Fine Structure Constant, Harvard University, https://www.phys.uconn.edu/icap2008/invited/icap2008-gabrielse.pdf

5. T. Quinn et.al, The BIPM measurements of the Newtonian constant of gravitation, G
https://royalsocietypublishing.org/doi/pdf/10.1098/rsta.2014.0032

Electron Mass and State Energy Levels Resulting from Photon-Photon Interaction

D.T. Froedge

V032522

This paper presents a summary of a number of papers representing an alternate view of the physics of particle structure and particle interaction [1-8]. The papers present an alternate interaction mechanism between photons and mass particles based on the probability of photons actually existing on the Feynman action paths. The theories are far from finished but there are clear indications that electric and gravitation aspects are the same phenomena, and it is clear that the value of Fine Structure Constant is the result of the interaction of photons with sufficient energy to form an electron. Progressively there has been a better understanding of the underlying mechanisms involved in the theory and thus, there are some points that are better presented here.

Abstract

This is a new perspective on fields and forces based on the Feynman action path view of quantum mechanics.

Although the probable Feynman paths are well known and have real effects, overlooked has been the probability that the particles are actually on these paths and have real effects. Presented here is the showing that the effect of these photons is responsible for both electric and gravitational interactions.

Richard Feynman developed the sum-over-histories view of QFT more than 70 years ago, showing that the path action a photon takes in going from one point to another is the sum of all possible paths. The most compelling proof of this is the delay represented by the Anomalous Magnetic Moment of the electron and the Aharonov–Bohm effect. The anomaly is the path delay due to the probability of being off the primary path. It has been measured and calculated to incredible precision and this makes the probability of these internal rotating photons being at a distance from the center of a particle undeniable.

Order of Presentation

Introduction

Gravitation

The Feynman path integral formulation of QED has implications beyond the path delays associated with the anomalous gyromagnetic ratio. That is that the photon action paths represent the probability that photons on action paths in the mass of particles, are actually on all the defined paths throughout space. The presence of these photons can account for the speed of light, and the forces of electric charge and gravitation [7].

From the primary postulate in "Vacuum Polarization, Gravitation, Charge, and the Speed of Light",[7] the ratio of the change in c to the observer's value is the potential energy to the total energy is:

$$\frac{\Delta c}{c_0} = \frac{\lambda_{PL}^2}{\lambda_P^2} \frac{2\lambda_P}{r} \equiv \frac{2Gm_1}{c^2 r} = \frac{\Delta\varepsilon}{\varepsilon_0} \tag{1}$$

Eq. (1), is a specific energy relation between the total invariant mass energy of the m_2 particle to the change of its mass energy due to the change in c generated by the Feynman photons generated by the sum of all the individual particles $m_1 = n m_P$ mass.

Δc is the difference between the value c at the observed position , and c_0, the value in the reference frame,. $\Delta c = c_0 - c$ Eq.(1), can be rewritten as

$$\frac{\Delta c}{c_0} = \frac{2\mu}{r} \rightarrow \frac{c}{c_0} = \left(1 - \frac{2\mu}{r}\right) \tag{2}$$

This is the Schapiro Time Delay measured change in c as the result of the presence of a local mass, found by the projection of the GR metric onto Minkowski flat space [9]. The theory is thus in physical agreement with General Relatively, in regard to the effect of gravity on c.

If there is a mass particle m_2, the specific energy ratio at the potential point the expression Eq.(1) is ,

$$\frac{\Delta c}{c_0} = \frac{2\mu}{r} = \frac{2Gm_1 m_2}{\left(m_2 c_0^2\right) r_1} \tag{3}$$

This is not the gravitational field potential, but the change in the total energy, $\Delta\varepsilon / \varepsilon_T$, or $\Delta(mc^2)/ mc^2$ as a function of r.

This equation defines the change in the velocity of light due to the probability of the Feynman photon density from the n protons of m_1 passing at a distance r .at the location of m_2

The change in c induces a change in the "specific" relativistic energy of $\varepsilon = m_2c^2$, which is the source of the gravitational potential energy. It is not the standard Gravitational or electric field potential, but an invariant energy potential. (See [7] for details on potentials)

$$\frac{\Delta c}{c_0} = \frac{pc_0 - pc}{pc_0} = \frac{\varepsilon_0 - \varepsilon_k}{\varepsilon_0} = \frac{\varepsilon_P}{\varepsilon_0} \tag{4}$$

This is the invariant energy of the system due to the invariant potential energy, or the "total" energy level on falling from infinity to r_1.

$$\varepsilon_0 = \varepsilon_k + \varepsilon_P \tag{5}$$

The energies of mass particle in a reduced velocity of light, is a specific energy function independent of the total energy. For this expression, if ε_0, is the total relativistic energy $\varepsilon_0 = hv = mc_0^2$, or the Hamiltonian of a mass, or photon in the altered c. ε_k is the kinetic energy of the photon $\varepsilon_k = mc_0c = p_0c$, and ε_p is the invariant potential energy or rest energy as a result of being in an altered value of c. (See: Appendix II, and the paper on photon wavefunctions, [1] for further discussion on this.)

Eq.(1), is not the gravitational field energy, but the total energy potential. In a conservative system it includes the sum of both the Kinetic and the invariant energy potential.

In a non-conservative system with the velocity of the m_2 particle in Eq.,(4) fixed, the static force is:

$$\Delta \varepsilon = \left(m_2c_0^2 - m_2c_0c \right) = \frac{Gm_1m_2}{r} \tag{6}$$

Eq.(6), is as noted in Eq.(5), a Hamiltonian expression.
Since force is $d\varepsilon / dr$ the force on the m_2 particle is then:

$$f = \frac{d\Delta\varepsilon}{dr} = -m_2c_0\frac{dc}{dr} = -\frac{Gm_1m_2}{r^2} \tag{7}$$

and is the proper gravitational force on the m_2 particle.

Mass of Electron - The First Energy Level in the Universe

The electron is the lowest possible value of invariant rest energy in the universe. It is the result of the binding of two photons that have sufficient energy to create a gradient in the radial index of refraction between the particles sufficient to overcome centrifugal effects, creating a stable invariant mass particle.

The energy of the photons necessary for that is determined by the ratio of the density of the Feynman photons in the universe to the self-generated density of the rotating photons. (See Appendix I, in regard to the nature of Photons.)

Mechanics of the Electron

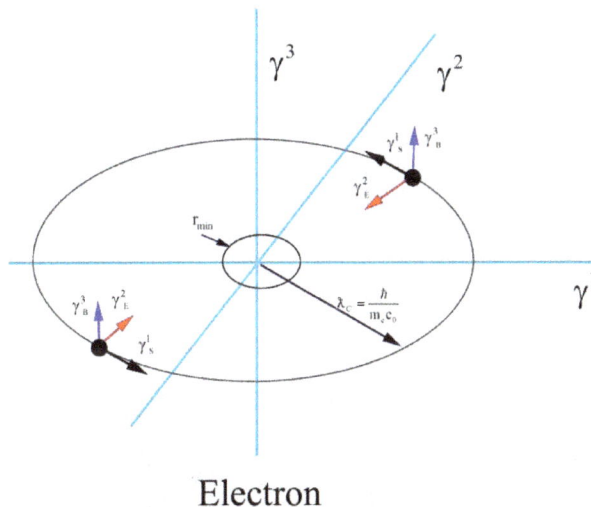

Electron

Fig.1. The vector orientation of the probability of location of two radially polarized photons rotating in the electron at the first most probable $L = \hbar$, orbit with r equal to the electron Compton radius. Red is the electric vector Blue is the magnetic vector and Black is the Poynting vector.

The electron is a composition particle composed of two photons, each with half the energy of the electron, revolving around the center of mass at a probable distance of the Compton radius. A photon passes or engages the probability of the other photon twice per revolution thus the frequency of photons passing the other photon is twice the rotation frequency. For a fixed point, there are two photon passing per revolution, which provides the proper Compton frequency for the electron. The Compton frequency is twice the rotation frequency of the photons.

From earlier papers, [2-9], noted in Eq.(1), the change in the velocity of light of a passing photon as the result of the probability of another photon passing at a perpendicular distance r is:

$$\frac{\Delta c}{c_0} = \frac{\lambda_{PL}^2}{\lambda_{ph}^2}\left(\frac{\lambda_{ph}}{r}\right) = \left(\frac{\lambda_{PL}^2}{\lambda_{ph}^2}\right)\left(\frac{\hbar}{m_{ph}cr}\right) \qquad (8)$$

(cross-section ratio) × Feynman photon probability density

The first term in brackets is change in the density of Feynman photons as the result of encountering the "ph" photon, and the second is the probability of the "ph" photon passing at a perpendicular distance r from the observed point. The momentum of the photon is $m_p c_0$, the Planck particle radius is λ_{PL}, and the Compton radius of the photon is λ_{Ph}.

If a photon is bound in orbit, at a frequency per revolution of v_1, an interloping photon encounters the universal background density of Feynman photons per second, [2-4], plus the density of rotating photon. This slows the velocity of light proportional to the frequency.

$$\frac{\Delta c}{c_0} = \frac{\lambda_{PL}^2}{\lambda_1^2}\left(\frac{\lambda_1}{r}v_{ph}\right) \qquad (9)$$

The other photon in the electron is not an interloping photon, but participating. The velocity of the photons is proportional to the density of Feynman photons in the universe, plus the density of encountering the other photon. The change is proportional to the probability of encountering the other photon, thus the "coincidence: of the probable encounter of the photons in an orbit.

The product of the probability of the photon densities, thus the "coincident" probability of collision is:

$$P_1P_2 = \frac{\lambda_1}{r}v_{ph}\frac{\lambda_2}{r}v_{ph} = v_{ph}^2\frac{\lambda_1}{r}\frac{\lambda_2}{r} \qquad (10)$$

132

The probability densities expressed here are for the Feynman photons of photons passing at a distance. The square of the frequencies v_{ph}^2 is the coincident collision frequency of the photons.

QM Perspective

From the perspective of Standard QM the two probabilities of photon location in Eq.(10), would be the wave equation amplitude. The product is not actually a particle location probability but the location density of the photon-photon collision. The density of this alters the mutual velocity of light of the orbiting photon.

The orbiting photon passes a point the repetition density is once per revolution, but since the other photon is also orbiting the encounter rate is twice the frequency of rotation, thus:

$$v_{ph} \rightarrow 2v_{ph} \qquad r = \lambda_{ph}/2 \qquad (11)$$

The corresponding equation, Eq.(31), for the change in c at the center of mass for the two photons becomes:

$$\frac{\Delta c}{c_0} = \rightarrow \left(\frac{\lambda_{PL} 2v_{ph}}{r} \right)^2 \qquad (12)$$

The numerator of this expression is the minimum radius of the orbiting photons, and the radius at which the centrifugal force equals the binding force of the two photons. (See Appendix IV Centrifugal force in electron)

The frequency of rotation is v_e thus the distance the photon travels at this distance in one second at the core of the electron c_e is $S = 2\pi(\lambda_{PL} 2v_{ph})v_e$. The ratio of this distance to the distance light moves, c_0 in one second is:

$$\frac{c}{c_0} = \frac{(\lambda_{PL} v_e) \times 2\pi v_e}{c_0} = \frac{\Re_F \omega_e}{c_0} \qquad (13)$$

The radial value $\mathfrak{R}$ is designated as the "Electron Energy Radius" since it represents the radius at which the potential energy of the two rotating photons is equal to the total energy The maximum of the potential energy can only be half the total energy at the point at which it is equal to the kinetic energy, thus we will define

$$\mathfrak{R}_0^2 = 2\lambda_{PL}^2 v_e^2 \tag{14}$$

This is the core radius of the electron it is an invariant quantity that defines the invariant mass. Two photons cannot actually be closer than this is because the kinetic energy of the photons p would be less than the invariant energy.

The frequency v_e in this expression is a pure number since it represents distance the photons rotate in the electron to the linear distance traveled in the same time

$$\frac{s_e}{s} = \frac{2\pi r}{c_0 t} v_e \tag{15}$$

Letting the Compton frequency of the Planck particle be $v_{PL} = c_0 / 2\pi D_{PL}$, the ratio of the velocity of light in the center of the electron to the free value in the universe is equal to the ratio of the coincidence frequency of photons in the electron to the Compton frequency of the Planck particle:

$$\frac{c_e}{c_0} = \frac{v_e}{v_{PL}} v_e \tag{16}$$

This is the condition for the formation of the electron.

From the vacuum polarization paper [7], the ratio of the velocity of light c_0 to the value at another point, c_e is inversely proportional to the ratio of the local density, to the Feynman photon density in the universe, n_f thus:

$$\frac{c_e}{c_0} = \frac{n_f}{n_e} \tag{17}$$

The density confronting the photon in the minimum orbit is the coincidence density, v_e^2 but as noted before, on each orbit the photon encounters the opposite photon twice, thus the density experienced by the orbiting photons is:

$$n_e = 2v_e^2 \qquad (18)$$

Putting this into Eq.(13), gives the value of c_e at the minimum possible orbit to be:

$$\frac{c_e}{c_0} = \frac{2\pi\lambda_{PL}v_e^2}{c_0} = \frac{n_f}{2v_e^2} \qquad (19)$$

Solving this for the number density of Feynman photons in the universe gives:

$$n_f = v_e^2 v_e^2 \frac{4\pi\lambda_{PL}}{c_0} = 1.575538679E + 38 \quad \text{photons sec}^{-1}\text{cm}^{-1} \qquad (20)$$

This Feynman photon density is within about a 7%, of the estimated vacuum density of Feynman photons in the universe based on the estimated mass in the universe, i.e. $n_f = 1.6928448E + 38$ photons sec^{-1}cm^{-1}, (see [7])

The Compton frequency and thus energy of the electron $\varepsilon = hv$ is then:

$$v_e = \sqrt[4]{\frac{c_0 n_f}{4\pi\lambda_{PL}}} \qquad (21)$$

Eq.(19), can also be expressed in terms of the Fine Structure Constant as defined by the energy values associated with the atomic state values in the next section.

$$\frac{c_e}{c_0} = \frac{2\pi\lambda_{PL}v_e^2}{c_0} = \frac{n_f}{2v_e^2} = \frac{\alpha}{\sqrt{2}} \qquad (22)$$

The state value of the two photons bound in the center of an electron Eq.(4), is then:

$$\frac{\Delta c}{c_0} = \frac{\Delta\varepsilon}{\varepsilon_T} = \left(1 - \frac{\alpha}{\sqrt{2}}\right) \qquad (23)$$

It is asserted that this is the minimum energy necessary for a photon capable of binding with an equal photon, forming an electron, and **this is the first invariant mass energy level of the universe.**

Although Eq.(16), and Eq.(23), establish the conditions for the mass of the electron, the minimum orbit is not a probable orbit for the electrons, for it is not a solution of the continuity relationship required as the solution of the sum of the wave equations of the two electrons.

The sums of the wave equations for two photons require geometry of a standing wave to have integral values of angular momentum. Over a given space there is an integral value of the magnitude of the sum of their momentum. i.e. $\lambda = \hbar / p$. [1]. That is if, $\psi = \psi_1 + \psi_2$, then:

$$\left| \gamma^\mu \frac{\partial \psi}{\psi \partial x} \right| \rightarrow \Delta x = \frac{np}{\hbar} \rightarrow \frac{1}{n} = \frac{\lambda}{p \Delta x} \tag{24}$$

Separating the integral values in Eq.(12), gives:

$$\frac{\Delta c}{c_0} = \rightarrow \left(\frac{\lambda_{PL} \, 2v_{ph}}{\lambda} \left[\frac{\lambda}{r} \right] \right)^2 \tag{25}$$

This has a solution when $c = \frac{1}{2} c_0$, thus at that radius $\lambda_{ph} \rightarrow \lambda_{ph} / 2 = \lambda_e$, and at that radius:

$$\frac{\Delta c}{c_0} = \left(\frac{\lambda_{PL} v_e}{\lambda_e} \left[\frac{1}{n} \right] \right)^2 = \frac{\Delta \varepsilon}{\varepsilon_T} \tag{26}$$

This is the state value of the photons in the electron. The following section will show the relation of this to α.

Note that the value of λ_e for the photon is:

$$\lambda_e = \frac{\hbar}{p_e} = \frac{\hbar}{m_e \dfrac{c_0}{2}} = \frac{\hbar}{p_0 / 2} \tag{27}$$

The h photons in the electron orbit now have an angular momentum of ½ h. The energy of the two photos $\varepsilon = v_e h$ is the energy required to bind the two photons together and we could define the radius $\mathfrak{R}_0 = \lambda_{PL} v_e$ as the Fermion radius.

Atomic Bound Particles

When two electron +/-engage the rotating planes of the opposite going photons come together since the planes represent the location of the highest index of refraction and align the photon planes. This causes a paring of opposite particles and is responsible for the Pauli Exclusion Principle

Note that the model of the atom presented here is not that of an electron orbiting a central particle or nucleus with a spherically symmetric radial attractive force.

The electrons can approach to radii of equal distance to the observation point and from that point each have an integral angular momentum, defined by the Compton radius $r = \lambda_e$. That is: the closest distance is when the Compton radii are in contact. This planar alignment gives the appearance of a spherical potential. The two photons in the electron rotate around their respective center of mass, but due to the minimizing effect on the index of refraction of the opposing velocity the rotation planes align in the same plane. The electrons can rotate around the center of mass of the two particles but the total energy and momentum in a fixed state is conserved, and the rotational orbit is not defined by centrifugal force.

The particle-particle interaction for the two particles is in principle exactly the same as for the two photons

For each electron, there are two Feynman photons, thus twice the number as for a photon and thus the frequency of the electron is the sum of the frequency of the two photons. The primary orbiting radius is at $r = \lambda_e$ where the velocity is ½ c, thus:

$$\lambda_e = \lambda_{ph} / 2 \qquad v_e = 2v_P \qquad m_1 + m_2 = m_e \qquad (28)$$

The change in the velocity in the proximity of an electron is now found identical to the composition of the two photons in Eq.(12), The expression for the velocity of light for interloping photons in proximity to the electron is then:

$$\frac{\Delta c}{c_0} = \frac{\lambda_{PL}^2}{\lambda_e^2} \left(\frac{\lambda_e}{r} v_e \right) \tag{29}$$

The other electron in an atom is not an interloping photon, but participating and the change in c for another internal photon is proportional to the "coincidence" probability of the encounter of the photons in orbit.

The coincident probability is the probability of orbiting photon encountering the oncoming photon density of the other photons. This increase in oncoming photon density as a function of r becomes a radial gradient in c binding the photons together.

The product of the probability of the photon densities, thus the coincident probability alters the velocity of the interaction photo.

$$P_1 P_2 = \frac{\lambda_e}{r} v_e \frac{\lambda_e}{r} v_e \tag{30}$$

The probability of the location of a moving photon is perpendicular to its velocity, thus orbiting photons encounter the probability of the other photon moving in an opposite direction. It is significant that Feynman's postulates regarding path directions that the path probability is in all directions.

The change in c for each of the photons from the two encountering $\pm$ electrons, as the result of the other is then:

$$\frac{\Delta c}{c_0} = \frac{\lambda_{PL}^2}{\lambda_e^2} \left(\frac{\lambda_e}{r_e} v_e \right) \left(\frac{\lambda_e}{r_e} v_e \right) \tag{31}$$

With the change in c being proportional to the inverse of the speed of light, this expression provides the radial binding energies for the photons in the electron. The radii r_1, and r_2 are the distance from an observation point to the respective particle For the electron the observation point is the center of mass with $r_1 = r_2$. The value of $\Delta c = c_0 - c$, thus the change in c is just the expression of the total energy plus the photon kinetic energy pc_0. The expression can be written as:

$$\frac{mc_0^2}{mc_0^2} = \frac{p_0 c}{mc^2} + \left(\frac{\lambda_{PL} v_{Ph}}{r}\right)^2 \tag{32}$$

The terms are recognizable, left to right as the ratio of the total energy to itself, the ratio of the kinetic energy to the total energy, and the ratio of the invariant potential energy to the total energy. This is the Hamiltonian for the bound photons, or:

$$m_e c_0^2 = p_0 c + V \tag{33}$$

The Lagrangian form is:

$$L = p_0 c - V = T - V \tag{34}$$

For further discussion see Appendix II and the section on photon energies in the paper, "The Dirac Equation and the Two Photon Model" [1].

The probability of the photon location exists at all radii, but as the radii of the photons goes down the binding energy goes up and the kinetic energy decreases due to the decrease in c. The limit for this is when the binding energy $\varepsilon_P = V$ or the invariant potential is equal to the kinetic energy $\varepsilon_K = T$ or.

$$p_0 c = \left(\frac{\lambda_{PL} v_e}{r}\right)^2 \tag{35}$$

The Lagrangian is the difference of these terms, and at the "minimum", (r_{min}), value of r in Eq.(29), requires the Hamiltonian potential expression to become:

$$\frac{mc_0^2}{mc_0^2} = \frac{p_0 c}{mc^2} + \left(\frac{\lambda_{PL} v_e}{r}\right)^2 \rightarrow 1 = \frac{1}{2} + \frac{1}{2}\left(\frac{\sqrt{2}\lambda_{PL} v_e}{r_{min}}\right)^2 \tag{36}$$

The equipartition of the energy at minimum r requires a minimum value to c to be $c = c_0 / 2$ at:

Rewriting Eq.(33), in terms of $\Delta c / c_0$:

$$\frac{\Delta c}{c_0} = \frac{1}{2}\left(\frac{\sqrt{2}\lambdabar_{PL}v_e}{r}\right)^2 = \frac{\Delta \varepsilon}{\varepsilon_T} \qquad (37)$$

This is the expression giving the value of the ratio of the change in invariant rest energy to the total energy as the particles revolve in a mutual orbit. At $r = \sqrt{2}\lambdabar_{PL}v_e$ the value of the term in parenthesis is equal to one and the velocity of light at that orbit is $c = c_0 / 2$

Eq. (37), expresses the relation between the total energy and the speed of light between two electrical particles. See Appendix III regarding the relation between the specific energy and the electrical forces.

Electron-Positron - Positronium Atom,

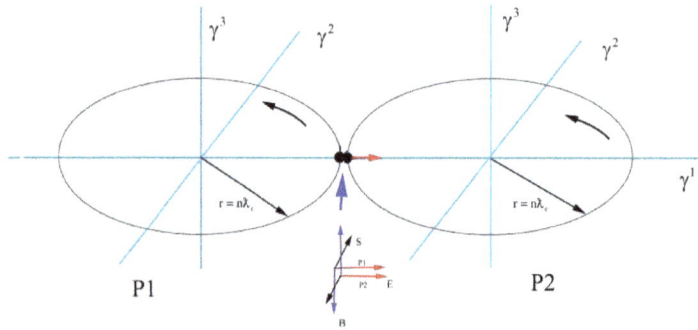

Fig. 2. The interaction of the Electron-Positron atomic system

Integral Geometric Constraints

Solutions to the electron wave equation, [1] require that the angular momentum in a geometrical configuration for the photons be integral values of $\hbar$. The equation expresses the continuity condition, in that the probability of location is a conserved quantity and is preserved when the angular momentum of the photon is integral as discussed in Eq. (24),

$$\gamma^\mu \frac{\partial \psi}{\psi \partial x} = \frac{np}{\hbar} = \frac{1}{\lambda} \rightarrow \lambda = \frac{\hbar}{np} = \frac{\hbar}{n\,mc} \qquad (38)$$

The space interval λ is an interval representing the ratio of $\hbar$ to the angular momentum $n\lambda = \hbar / p$ Frim the electron wave equation the value of $\lambdabar$ for the electron is the Compton radius, $\lambdabar_e = \hbar / m_e c_0$.or $\ell = m_e c_0 \lambdabar_e$

Since this is an integral value it can be separated as a factor in Eq. (37), as an integral term. For the Compton radius of the electron this is:

$$\frac{\Delta c}{c_0} = \frac{1}{2}\left(\frac{\sqrt{2}\lambda_{PL}v_e}{\lambda_e}\left[\frac{\lambda_e}{r}\right]\right)^2 \rightarrow \frac{1}{2}\left(\frac{\sqrt{2}\lambda_{PL}v_e}{\lambda_e}\right)^2 \tag{39}$$

At that radius $c = c_0/2$ the wavelength of the photon is reduced to ½, making the radius equivalent to the Compton radius of the electron and the expression can is that of a state value of the energy an atom.

State Energy Levels

Solutions for state systems for the Schrodinger and the Dirac equation exist when the photon standing wave wavefunctions have integral values in a particular geometrical configuration. This is the result of the photon location probability defined by its wavefunction having continuity as discussed for Eq. (26).

Setting the angular momentum in Eq. (39), to be $r = n\lambda_e$, or $\ell = n\hbar$ The results for the invariant state energy, Eq. (4), of an atomic system is:

$$\frac{\Delta c}{c_0} = \frac{1}{2}\left(\frac{\sqrt{2}\lambda_{PL}v_e}{\lambda_e}\left[\frac{1}{n}\right]\right)^2 = \frac{1}{2}\left(\alpha\left[\frac{1}{n}\right]\right)^2 \tag{40}$$

It has been shown in earlier papers, that the term in parenthesis is to a high accuracy equal to the "Fine Structure Constant"[5],[6] thus the value of the binding energy is just the ratio of the Rydberg energy to the total energy of the electron.

$$\alpha = \frac{\sqrt{2}\lambda_{PL}v_e}{\lambda_e} = 1/137.0359997100 \tag{41}$$

Notation change (See Appendix I p16)

$$\frac{\Delta c}{c_0} = \frac{\Delta\varepsilon}{\varepsilon_0} = \frac{\alpha^2}{2n^2} = \frac{1}{m_e c^2}\frac{R}{n^2} \tag{42}$$

This is the energy ratio of the electron mass energy, to the Rydberg energy, and the energy difference between two energy states in the positronium atom is:

$$\Delta\varepsilon_2 - \Delta\varepsilon = R\left(\frac{1}{n_2^2} - \frac{1}{n_1^2}\right) \tag{43}$$

The well-known ionizing spectral energy of Positronium is half the Rydberg energy:

$$\hbar\omega = \frac{R}{2} \tag{44}$$

It is asserted that the energy of the first state of Positronium Eq.(42), is the Rydberg energy, but the spectral energy observed from Positronium has only half the Rydberg of this.

It is well known that the Schrodinger Equation, defining the atomic energy levels in an atomic system is a geometrical solution to a standing wave. A single photon cannot be a conjugate phase photon to itself, and it is asserted that each level must have two conjugate photons.

Thus there are two equal conjugate phase photons emitted when the election falls into the lowest state. It is presumed that the second photon is scattered and generally lost to kinetic energy, (See Positronium Appendix V,)

Conclusion:

This paper presents physics with an alternate mechanism of the force between particle and the structure of particles. It is believed to be a gateway to the understanding and the mass of heavier particles.

This paper presents the results of a number of papers identified in the references. As there has been a better understanding of the underlying mechanisms, there are some points that are better clarified here, and some that still need work. The issue of charge sign is a straightforward application of spin and Lorentz photon interaction, but is left to a later paper.

The basic theory is far off the standard track, and thus far no one in the current physics community, sufficiently skilled to understand the implications, has paid much attention. It is not wrong however, and at some point, it will come to the forefront.

References:

More extensive lists of on referenced papers, are found in the references 1-6 by Author. Some of the papers 1, 2 are a little dated and need some revisions.

1. DT Froedge, The Dirac Equation and the Two Photon Model of the Electron revised, April 2021, DOI:10.13140/ RG.2.2.19095.70564, https://www.researchgate.net/ publication/350922403

2. DT Froedge, The Connection between Electric Charge, Gravitation, and the Feynman Sum over All Histories View of Quantum Electrodynamics, April 2020 Conference: APS April. 18-21, 2020
Washington,DC.https://absuploads.aps.org/presentation. cfm?pid=18355
https://www.researchgate.net/publication/341310206

3. DT Froedge, A Quantum Theory Conjecture on the Origin of Gravitational and Electric Particle Interaction, December 2019, DOI: 10.13140/RG.2.2.29097.54884
https://www.researchgate.net/publication/337826826

4. DT Froedge, The Electron as a Composition of Two Vacuum Polarization Confined Photons, April 2021, DOI: 10.13140/RG.2.2.18971.18722 , https://www.researchgate.net/ publication/350740864

5. DT Froedge, The Gravitational Constant to Eleven Significant Digits, March 2020, DOI: 10.13140/RG.2.2.32159.38564
https://www.researchgate.net/publication/339943651

6. DT Froedge, The The Fine Structure Constant from the Feynman Path Integrals, March 2021, DOI:10.13140/RG.2.2.12979.55846, https://www.researchgate.net/publication/350188862

7. DT Froedge, Vacuum Polarization, Gravitation, Charge, and the Speed of Light, Sept. 2021
DOI: 10.13140/RG.2.2.15619.22569, https://www.researchgate.net/publication/354474157

8. Dirac, P. A. M. (1934). "Discussion of the infinite distribution of electrons in the theory of the Positron". Cambridge Phil. Soc. 30 (2): 150–163. Bibcode:1934PCPS...30..150D. doi:10.1017/S030500410001656X.

9. Roger Blandford, Kip S. Thorne, Applications of Classical Physics, (in preparation, 2004), Chapter 26
http://pmaweb.caltech.edu/Courses/ph136/yr2012/1227.1.K.pdf

10. S. Errede, Lecture Notes 16,The Structure Of Space-Time, Sept 2015,
http://web.hep.uiuc.edu/home/serrede/P436/Lecture_Notes/P436_Lect_16.pdf

Appendix I

The Photon

The photon is proposed to be a Planck particle with an area $\lambda_{PL}^2 = \mu\lambda = G\hbar/c^3$ revolving at the photon frequency ω such that, the Compton radius, is equivalent to the distance from the center that light travels in one cycle. The rotation of the photon is either right or left handed around the velocity vector. The probability of the particle location oscillates linearly back and forth through the particle center as the particle rotates. Beyond the Compton radius, the photon's probability exists, but only for that of the bare Planck particle with a density repetition rate equal to the Compton frequency. Beyond the core Compton radius of the photon there is a passing probable density of the bare Planck particle. The Frequency of the bare Planck particle is $\omega_{PL} = c_0 / \lambda_{PL} \approx 1.8E+43 \, \mathrm{rad \, sec^{-1}}$

Photons in the electron have orbital frequency matching thee Compton frequency, thus a constant radial polarization.

The probability of existence of the photons beyond the Compton radius is evidenced by the Anomalous Magnetic Moment of the electron, the Aharonov–Bohm effect, and from this development the existence of gravitation and charge. Beyond one wavelength however, the particles are so small, 10^{-66} cm², that the probability passes through dense mass without delay or scatter. This property is responsible for the observation that gravitational effects are not shielded. The actual effect, of the photons is to alter the speed of light, and change the relativistic energy $\varepsilon = mc^2$. For purposes here the mass of a photon is considered to be $m = p / c_0$, and the change induced by the velocity and density of the Feynman photons is:

$$\frac{\Delta c}{c_0} = \frac{\Delta \varepsilon}{\varepsilon} = \frac{\text{Change in } \varepsilon = mc^2}{\text{initial value of } \varepsilon = mc^2} \tag{45}$$

Appendix II

The Photon wavefunction

Excerpts from "The Dirac Equation and the Two Photon Model of the Electron" [1]

For a photon there is a wavefunction that expresses the probability of location as:

$$\psi = e^{i\left(\frac{mc_1}{h}\hat{k}gx - \frac{m_1 c_0}{h}ct\right)} \tag{46}$$

The invariant energy of a photon is then:

$$E_I = hc \left| \gamma^\mu \frac{\partial}{\psi \partial x^\mu} e^{i\left(\frac{mc_1}{h}\hat{k}gx - \frac{m_1 c_0}{h}ct\right)} \right| = 0 \tag{47}$$

The magnitude is the square root of the product of the four-vector times its conjugate.

If $c_1 = c_0$ the value of E_I vanishes thus there is no invariant rest energy, but if the location of the photon moves into a higher index of refraction such as piece of glass where $c_1 \neq c_0$ then the product or square of the four vectors is not zero, but a Lorentz invariant scalar [10], and the photon has not only relativist kinetic energy, but also a scalar invariant rest energy. This invariant rest energy creates rest mass equivalent in the rest frame of the glass. When it leaves the glass, the energy is again, pure relativistic energy, thus the glass has by slowing the velocity temporarily changed some of the energy from relativistic to invariant.

Three Parts of the Energy of a Photon

From the above discussion, the components of the total energy are; the invariant or a rest frame energy E_I, the kinetic energy E_K, and the total energy, E_T, specifically this is:

$$E_I = c_0 h |p^\mu| = c_0 h \left| \gamma^\mu \frac{\partial \psi}{\psi \partial x} \right| \qquad E_K = c_0 h \left| \gamma^k \frac{\partial \psi}{\psi \partial x} \right| \qquad E_T = c_0 h \left| \gamma^0 \frac{\partial \psi}{\psi \partial x} \right| \quad (48)$$

The difference of the kinetic and total energy is the Lorentz invariant scalar rest mass, or the binding potential energy.

Appendix III

Electric Force

The expression for the energy relation between electrical particles in Eq.(37), may be construed as the effect of an electrical field.

To find the forces between charged particles the expression for the electron interaction between two particles can be written with the observation location not at the center of mass.

$$\frac{\Delta c}{c_0} = \frac{1}{2} \left(\frac{\sqrt{2} \lambda_{PL} v_e}{\lambda_e} \left[\frac{\lambda_e}{r} \right] \right)^2 \rightarrow \frac{1}{2} \left(\frac{\alpha \lambda_e}{r_1} \right) \left(\frac{\alpha \lambda_e}{r_2} \right) \quad (49)$$

The expression in Eq.(49), is the change in the specific energy of the two particles energy. It can be expressed as the coordinate differential energy if we set the observation point at the minimum radius of one of the particles $r_1 = \alpha \lambda_e$. The other r_2 is then the spatial separation. Writing this out gives:

$$\frac{\Delta c}{c_0} = \frac{1}{2}\left(\frac{\alpha \lambda_e}{r_2}\right) = \frac{1}{2}\left(\frac{\alpha c \lambda}{m_e c^2 r_2}\right) \tag{50}$$

Where r_2 is the distance from particle 1 to particle 2

$$2\left(m_e c_0^2 - m_e c_0 c\right) = \frac{Q^2}{r} \tag{51}$$

The right of this expression is the nominal electric energy, the left side is the total change in the energy of both particles, and thus the electric energy derives from the mass energy of both particles. And the force between the particles is:

$$f = \left(m_1 + m_2\right)c_0 \frac{dc}{dr} = \frac{Q^2}{r^2} \tag{52}$$

Appendix IV

Centrifugal force in electron

At the minimum radius of the electron in Eq.(12), the centrifugal force is equal to the gradient n the velocity of light that binds the photons together.
The equation is

$$\frac{\Delta c}{c_0} = \frac{\Delta \varepsilon}{m_{ph} c_0^2} = \rightarrow \left(\frac{\lambda_{PL} 2v_{ph}}{r_1}\right)^2 \left(\frac{\lambda_{PL} 2v_{ph}}{r_2}\right) \tag{53}$$

Let r_1 be equal to its minimum value $\lambda_{PL} 2v_{ph}$ the expression becomes the differential energy between the particles as a function of r_2

$$\frac{\Delta\varepsilon}{m_{ph}c_0^2} = \rightarrow \left(\frac{\lambda_{PL} 2v_{ph}}{r}\right) \tag{54}$$

Taking the differential of this expression with respect to r_2, which is the gradient in the energy, gives the force on the photon.

$$f = \frac{d\Delta\varepsilon}{dr} = \rightarrow \left(\frac{\lambda_{PL} 2v_{ph}}{r}\right) m_{ph}c_0^2 \tag{55}$$

If this is set equal to the centrifugal force on the electron $f_c = m_{ph}c^2/r$, the value of r_2 is:

$$r_2 = \lambda_{PL} 2v_{ph} = \lambda_{PL} v_e \tag{56}$$

Thus when the particles arrive at this radius the binding forces match the centrifugal forces and the particle is created,

Appendix V

Fundamental Issues

Spin

The orbiting electrons have integral half spin angular momentum ℓ, in a two particle system the attraction between the two particles is due to their opposite going photons, and therefore the B fields are anti-aligned. If a positron is subjected to a magnetic field there is created a preferential direction for its orientation. It can flip to the opposite orientation in another state, creating the observed spectral splitting. The value of this is apparent in the wave equation solutions, [1].

Positronium State Values

It would appear in Eq.(43), that the energy value for the state is twice the proper value as noted in Eq.(44), since the spectral energy levels for positronium is ½ R. Each photon orbiting in the electrons has a spectral energy equal to ½ R that is radiated away when the particles come together. It has been previously assumed that the spectral energy is the total energy of the state however; it appears from the preceding presentation that the state energy level is twice the spectral energy.

The Schrodinger Equation defines geometrical energy levels for standing waves as the same as described here, but ignores the fact that a standing wave represents two opposite conjugate phase waves. These states are more clearly defined in the two photon model [1], [4].

It is asserted here that the levels nominally calculated via the Bohr and Schrodinger represents represent only half the state energy levels and the other photon is scattered or lost to kinetic energy in the radiation process, and ignored.

Forces

Since Newton developed the concept of force, it has been the inspiration for a multiplicity of continuous field theories. The defined fields are generally infinitely divisible and lead to infinities with no explanation.

In the creation of a field description of a physical phenomenon, starting with the Newton concept of force the motions is expressed, in terms of physical parameters. The next step is to define a field imbued with an abundance of invisible energy that transfers the action between particles. Leibniz, refereed to gravitation as a return to occult quantities, and in reality there is magic and spooky action at a distance involved.

It is asserted that that the assignment of energy to a continuous field is a pervasive error that that cannot be fixed, and has led to the un-reconcilable differences in Gravitational and Electrical theory. This paper expresses a theory devoid of the phenomena of fields.

Vacuum Polarization, Gravitation, Charge, and the Speed of Light

D.T. Froedge

V090921

Author Comment:

This paper and the series of papers associated with, introduce a new basic view of physics providing originating mechanisms for rest mass, gravitation, electric forces, and the velocity of light in the universe. The originating mechanism of the mass of the electron is established. The Fine Structure Constant is calculated in terms of the physical constants, within the measured experimental error bars. The relation between Gravitational constant and the fine structure is calculated accurately without fudge factors.

These ideas are not going to come into the mainstream of physics for at least many years after my life, not likely anyone will pay much attention, but I believe it is the starting point of the next revolution, and I have done the best I can to elucidate it.

Abstract

The concept of mass and particle interaction by way of force and continuous fields, has persisted since Faraday's time, and is still the prevalent theoretical mechanism in spite of the problems the infinities and the impossible energy densities. The concept of the Vacuum Polarization has also been known for almost a century and is the origin of the mechanism creating pair production for electrons, and setting the speed of light [22].

The purpose of this paper is to present a plausible alternate mechanism for the electric and gravitational forces, define an alternate mechanism for the vacuum polarization and the causation of the speed of light. This same mechanism, the presence of the probability of Feynman action path photons, creates the forces in atoms that bind electrons and atoms together.

Feynman has well illustrated that a photon going from one point to another, takes an infinite number of possible action paths. The probability of taking these paths throughout space is measurable and displayed by measurements of the Anomalous Gyromagnetic Ratio, g_A. The calculation and measurement of these path delays of electron loop probability, give a certainty that a moving photon exists not only on its classical path inside the Compton orbit, but its probability exists, and has a substantial effect throughout space. It is to be shown that this probability density creates the effect of charge, mass, and gravitation.

It is proposed that there is neither a Gravitational nor an Electric field energy density. The concept of electric and gravitational fields is useful in calculating potential energy and motion in atomic system and gravitational dynamics, but otherwise is a conceptual mistake. The change in c induced by the presence of the Feynman photon probability is shown to create the effect of both phenomena, and in fact are one in the same. (See Endnote 1)

This paper addressee primarily, time independent static electric and gravitational effects. The dynamic photon-photon interactions are better addressed by the continuity of the time dependent photon probability density expressed by the wave equation, and the photon four-momentum. The single photon wavefunction represents the time dependent continuity of the sum of the location probability, and the interaction of these wavefunctions creates a time independent Lorentz invariant standing wave solutions that represent mass particles. These are the standing waves that are the solutions to the Schrodinger equation .The dynamic interaction of photons requires the formalism of the photon wavefunction, and the attendant four-momentum that are developed in, "The Dirac Equation and the Two Photon Model of the Electron" [1]

Vacuum Polarization

It is presented here that "Vacuum Polarization" is not a field at all but the presence of the probability density flux of Feynman action path photon, generated by the internal action paths of photons residing in mass particles. The Vacuum Polarization of Dirac, Schwinger, et.al, is cast as a probability density of Planck particles, having no energy, no scatter or interaction with mass, but sets both rest mass and potential energy for the interaction of particles as the result of a change in c.

Main

Although ignored by most; if the Feynman action paths do exist, there is the probability of the presence of Feynman action path photons being elsewhere in space. It is asserted that the presence of Feynman photons generated by the internal motion of photons in mass particles, alter the speed of light, and change the energy of other mass particles mass via $E = mc^2$. The same change in the relativistic energy as a result of the change in the velocity of light generates the potential energy difference for both charge and gravitation. The spatial potential energy between mass particles is not generated by magic but by the interaction of the probability density of the Planck size photons.

Primary Postulate

The change in the speed of light moving thru a density of Feynman photons generated by mass particles is:

$$\frac{\Delta c}{c_0} = \sum_n \frac{\lambda_{PL}^2}{\lambda^2}\left(\frac{2\hbar}{mcr}\right) \quad \frac{\text{additional hits per second of travel}}{\text{misses per second of travel}} \qquad (1)$$

The ratio of the change in the velocity of light to c_o is equal the ratio of additional hit to misses of a Planck particles per second.

The first term on the right of this equation is the ratio of hit to miss on passing within the Compton radius of another photon, the second term is the probability density of the photon passing there.

Topics;

The Interaction Mechanism
Units and Concept
The Photon

Gravitation

Mass Induced Change in the Local Velocity of Light
Gravitational Potential Energy
Vacuum Polarization and the Mechanism Setting Speed of Light in the Universe

Electric Field and Charge

The Electron Binding Density and Vacuum polarization
Coincident Probability Density (collision density) of Interacting Photons
Structure of the Electron
    The Electron and the Fine Structure Constant
    Two Particle Energy Potential
    Atomic Energy States, A New Atom
$E = mc^2$ in a Variable Speed of Light
Relation of Schwinger Vacuum Polarization Electric Field Density and the Photon Flux Density
Conclusion
References
Appendix I *Calculated Fine Structure Constant*
Appendix II *Compton Radius Adjustment*
Endnotes

The Interaction Mechanism

The speed of light is set by the ratio of the probable hit to the probable miss per second of the Feynman photons. The Feynman photons being the Planck size photons generated from the action paths of the internal motion onside the Planck radius of mass particles.

Each, half spin boson particle generates the probability two photons and by the same concept of the action paths, there is a probability density of location outside their classic path of $\hbar / L$. The density is thus inversely proportional to the angular momentum. The probability density as a function of r is found to be:

$$\text{Spin } 1/2 \text{ boson} \qquad P = \frac{2\lambdabar}{r} = \frac{\hbar}{mcr/2} \qquad (2)$$

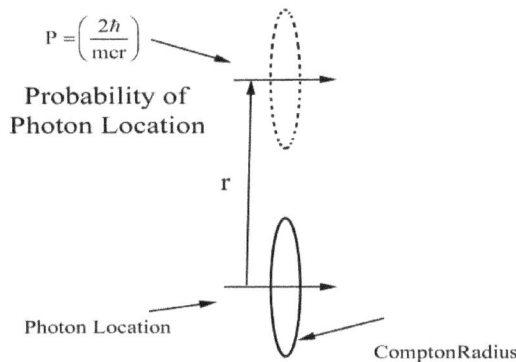

$$P = \left(\frac{2\hbar}{mcr} \right)$$

Probability of
Photon Location

r

Photon Location

Compton Radius

Fig. 1. The probability of a photon being at a distance from the center of its Compton radius.

For a mass this density is the probability of a photon in its Compton radius will be moving in the same direction at a distance of r from the center of the Compton radius. The energy and angular momentum of the particle is quite definite, but the probability of where it is located from its center is this function of r. This is the probability density of a photon actually being on the probable action paths defined by Feynman. Feynman illustrated that the paths go in all directions and may be forward or reversed in time, but the pre-

155

sumption is that the probability of the action paths at a distance are parallel to the photon velocity.

The Δc per photon

The change in the velocity of light per single photon encountered per second shown in Eq.(1), is:

$$\frac{\lambda_{PL}^2}{\lambda_C^2} \tag{3}$$

This is the area ratio of the core Planck particle $\lambda_{PL}^2 = \mu\lambda$, in the photon, to the Compton area λ_e^2, or is the ratio of the hit to miss of a Planck size particle passing within the Compton radius of the photon. If n is the number density per unit area passing a point then the total change is:

$$\Delta c = \Delta c_n\, n \tag{4}$$

The mechanism of interaction of the Feynman photons with mass is not a kinetic transfer but is a change in c altering the relativistic energy relation, $\Delta\varepsilon = \Delta(mc^2)$. The differential in relativistic rest energy will be shown to provide the potential energy for both gravitational and electric interaction.

Historical Note

The concept of the nonlinear aspects of the electromagnetic field producing electron–positron pairs was first discussed by Dirac and Heisenberg in the 1930's, and the effects were calculated R. Serber and E. A. Uehling, [6], [7], [8], [9], [10], [11]. Later the values of the energy densities of the Vacuum Polarization necessary to create pair production were developed by Schwinger [12]. The history of the concept is well known in the physics community.

Units and Concept

The probability density of Feynman photons generated by a Spin ½ mass particle is equivalent to 2 photons distributed throughout space. This is easily understood for the electron which consists of only two orbiting photons. Other bosons must have the same intrinsic density that is generated by a sum of internal motion. The

Feynman photons all have the same Planck size and velocity, but the density is dependent on the mass of the generating particle.

For collections of mass particles the directions of the Feynman photons are random thus leaving gravitation being the result of the sum of random photon motion throughout space. For electrons the spin angular momentum of the two internal rotating photons are polarized and align and thus the vector interaction predominates.

The most minimal change in $\Delta c / c$ is the result of a probability density of intersecting an increase in Planck particles of one per second. For a photon passing within the Compton radius of one proton generated photon per second is:

$$\frac{\lambda_{PL}^2}{\lambda_P^2} = 5.9059571E - 39 \tag{5}$$

The Photon

The photon is proposed to be a Planck particle revolving at the photon frequency ω such that, the Compton radius, is equivalent to the distance from the center that light travels in one cycle. The probability of the particle location oscillates linearly back and forth as the particle rotates at is frequency. Beyond the Compton radius, the photons probability exists, but only for that of the bare Planck particle with a repetition rate equal to the frequency, but no spin nor Compton radius.

The probability of existence beyond the Compton radius is evidenced by the Anomalous Gyromagnetic Ratio and the effect on other photons. Beyond one wavelength however, the particles are so small, 10^{-66} cm^2, that the probability passes through dense mass without delay or scatter. This property is responsible for the observation that gravitational effects are not shielded. The effect, of a change in photon density is to alter the speed of light, and change the relativistic energy. For purposes here the mass of a photon is considered to be $m = p / c_0$.

Gravitation

Mass Induced Change in the Local Velocity of Light and Gravitation

From Eq.(16), a change in the value of c as the result of the local probability of Feynman photons is:

$$\frac{\Delta c}{c_0} = \left(\frac{\lambda_{PL}^2}{\lambda_C^2} \frac{2\lambda}{r}\right) = \frac{2\mu\lambda}{\lambda r} = \frac{2\mu}{r} = \frac{2Gm_1m_2}{(m_2c_0^2)r} \tag{6}$$

Noting that $\Delta c = c_0 - c$, the third relation in this can be written:

$$\frac{c}{c_0} = \left(1 - \frac{2\mu}{r}\right) \tag{7}$$

This is a known relation between the speed of light and gravitational mass. It has been measured by the Schapiro effect [16], and theorized from GR by the projection of the gravitational tensor onto Minkowski four-space by Blandford [17]. The value is thus in agreement with experiment. It can also be viewed from another perspective as providing the source of gravitational potential energy. (See Endnote 2)

Gravitational Potential Energy

The fourth term in Eq.(6), is the ratio of the specific energy, to unit mass of a test particle thus the change in c provides the source of gravitational potential energy.

With c being a function of the Feynman photon density generated by mass particles the mystery of gravitational energy potential can be better understood. It is easy to deduce the reciprocity between the relativistic mass energy and the velocity of light.

Leibniz's accusation of Newton's of introducing 'occult' forces because, gravity acts at a distance as if by magic. It is here asserted that Gravity is not a mysterious force, but a force generated by a probability density of photons produced by the mass particles existing on the Feynman action paths altering the speed of light near massive objects.

Putting a **test particle** m_2 in both sides of Eq.(6), shows the "specific energy" or energy per unit mass of the gravitational energy potential for a mass particle 2 to be:

$$m_2 c_0^2 \frac{\Delta c}{c_0} = \frac{2 G m_1 m_2}{r} \tag{8}$$

Noting that, $\Delta c = c_0 - c$ Eq.(8), can be stated as a change in the relativistic energy.

$$\Delta \varepsilon = \left(m_2 c_0^2 - m_2 c_0 c \right) = \frac{2 G m_1 m_2}{r} \tag{9}$$

The right side of this expression is the relativistic energy change of m_2 and the left is the total relativistic energy change of m_2 entering the gravitational potential, thus the specific energy per unit mass, and in a conservative system it includes both the potential and kinetic energy.

$$K_e + \frac{G m_1 m_2}{r} \tag{10}$$

It is concluded that **gravitational energy is equal to the change the relativistic rest energy** $(m_2 c_0 c)$ of a mass particle, as a result of a change in c, and Δc is the source of the observed gravitational potential energy. This casts the expression $\Delta \varepsilon$ as a Hamiltonian.

Vacuum Polarization and Gravitational set Speed of Light in the Universe

It is asserted that the speed of light in the universe is set by the probability density of Feynman photons generated by the mass particles. This sea of probable photons generated by the particle masses provides the limit to the velocity of light.

In an empty universe the velocity of light would be near infinite, (see Endnote 3), but the probability of collision with other photons, reduces the speed to finite values.

Employing the first postulate, Eq.(1), and noting that, as more and of the mass of the universe is included the calculation, $\Delta c = (c_0 - c)/c_0 \to 0$, the ratio of c to c_0 approaches one, and Eq.(1), becomes:

$$c = c_0 \sum_n \frac{\lambda_{PL}^2}{\lambda^2}\left(\frac{2\hbar}{mcr}\right) \tag{11}$$

From estimates of the distribution and mass of the universe, the number density of Feynman photons in universe as a whole can be estimated.

Most of the mass in the universe consists of proton mass particles and the most significant density contributor of Feynman photons. By summing Eq.(11), over the entire universe, setting it equal to the velocity of light, and simplifying the expression, it becomes:

$$c = c_0 \left(\frac{\lambda_{PL}^2}{\lambda_n^2} \sum_n \frac{2\lambda_n}{r_n}\right) = c_0 \left(\sum_n \frac{2Gm_P}{c_0^2 r_n}\right) \tag{12}$$

From an estimate of the mass in the universe by D. Valev [13], which is the same as setting the radius of the visible universe as the Schwarzschild radius, the relation between the mass and the radius in a flat universe is:

$$R = \frac{2Gm}{c^2} \qquad \text{or} \qquad \frac{2Gnm_P}{c^2(R)} = 1 \tag{13}$$

With the radius of the universe determined to be.
$R = 1.2961570E + 28$ cm, $[15]$, the number of equivalent masses is:

$$n_p = \frac{R}{2\mu_p} \quad = \quad 5.2176112E + 79 \tag{14}$$

From the presumption that the mass in the universe is primarily made up of protons and neutrons with approximately the same mass m_p. This sum can be integrated over the universe giving:

$$c = c_0 \left(\sum_n \frac{2Gm_p}{c_0^2 r_n} \right) = \int_V \sigma \frac{dv}{r} \rightarrow c_v \left(\frac{2nGm_p}{c^2 R} \right) \tag{15}$$

Comparing Eq.(12), and Eq.(13), the value of the ratio of c/c_0 for the universe is:

$$\frac{c}{c_0} = \left(\frac{\lambda_{PL}^2}{\lambda_p^2} \times \sum_n \frac{2\lambda_P}{r_n} \right) = 1 \tag{16}$$

The first term in this expression is just the ratio of the Planck area to the proton Compton area, and is the change in c due to the interaction with a single Feynman proton photon density $\lambda_{PL}^2 / \lambda_p^2$. The values for the Planck particle are:

$$\lambda_{PL} = 1.61624e\text{-}33 \text{ cm} \qquad \lambda_{PL}^2 = 8.2067e\text{-}66 \text{ cm}^2 \tag{17}$$

The reciprocal of this must be the number density of Feynman photons in the universe. n_f, can then be calculated from Eq.(16), as the reciprocal:

$$\frac{c}{c_0} = 1 = \frac{\lambda_{PL}^2}{\lambda_{Prot}^2} \times n_f = \frac{1}{1.6932057E + 38} \times n_f \tag{18}$$

The Feynman photon flux density n_f, in the universe is then estimated as:

$$n_f = 1.6932057E + 38 \tag{19}$$

161

This is the number density, n_f, of Planck particles per cm² sec, generated by protons passing throughout the universe.

This flux constitutes what has been known as the "Vacuum Polarization". Because of Feynman photon cross-section of the Planck particle, E-66 cm² there is no extractable, detectable or measurable energy content. The photon density affects the speed of light which in turn affects particle interaction, but the photons have no ascertainable energy content. **This contrasts with the consistency issues between quantum electrodynamics, and Lorentz covariance,** which that suggest energy density for the vacuum polarization electric energy to be greater than, 10^{123} ergs/cm² ,[14], [15]. This density of energy is obviously an absurd and undoubtedly erroneous result. (See Endnote 1)

Electric Field and Charge

The Electron Binding Density and Vacuum polarization

Described earlier is the illustration that a change in the flux density of oncoming photons alters the velocity of a moving photon. If two photons are moving in the same particle the concurrence of their densities alters their mutual speed.

For the two photon model of the electron, postulated in [1], & [4], the photon density encountered by one of the photons, in addition to the space vacuum value, includes the density provided by the other revolving photon. Although the photons in the electron are orbiting in the same direction Feynman action paths are sums over all directions and thus the photons in the Compton orbit encounters the same opposing probability density. The densities are a function of the angular momentum as noted above, and thus provide a radial gradient in the index of refraction in the electron. This Gradient in the index of refraction is not a force but induces circular orbits that bind the photons in the electron [22]. It can be compared to the force on a charge in an electric provided by the Schwinger Critical value. (See section comparing relation)

Self-Generating Index of Refraction in the Electron

It has been shown [1]. [2], that electrons are composed of two polarized photons in orbit in an electron.

From Eq.2, the probability density of a photon P_1 passing at a distance r from its center is:

$$P_1 = \frac{2\lambda_1}{r} \tag{20}$$

If the photon is orbiting in the electron, at a frequency per revolution of v, an interloping photon would encounter the background density in photons per second, plus the density of the internal photon at its frequency. Thus the probability density of the orbiting photon is multiplied by its rotation rate, and from Eq.(1), the change in c for the interloping photon would be:

$$\frac{\Delta c}{c_0} = \frac{\lambda_{PL}^2}{\lambda_1^2} \frac{2\lambda_1}{r} v_1 \tag{21}$$

The other photon in the electron is not an interloping photon, but also orbiting, and the change in c is the probability of one of the photons being in the same place as the other. This is the coincident probability or the product of the photon densities.

$$P_1 P_2 = \frac{\lambda_1}{r} v_1 \frac{\lambda_e}{r} v_1 \tag{22}$$

The change in c for each of the photons as the result of the other is:

$$\frac{\Delta c}{c_0} = \frac{\lambda_{PL}^2}{\lambda_1^2} \frac{2\lambda_1}{r_1} v_1 \frac{2\lambda_2}{r_2} v_2 \tag{23}$$

This is the coincident collision density of the interacting photons.

The frequency is the Compton frequency $v_1 = c / 2\pi\lambda_1$, and the energy of each of these photons composing the electron. The value of the Compton frequency of the photons are half the Compton frequency of the electron, thus Eq.(23), can be:

$$\frac{\Delta c}{c_0} = \frac{\lambda_{PL}^2 v_e^2}{\lambda_e^2} \left(\frac{\lambda_e}{r_1} \frac{\lambda_e}{r_2} \right) \tag{24}$$

(Previous error corrected. See Endnote 4)

From the Born's* QM, view, P would be the probability amplitude, and the coincident probability would correspond to the square or wavefunction probability density. Here, it is the probability density of the location of a coincident event, thus the probability density of the location of opposing photons.

The change in their mutual speed of light is then the probability of a hit times the change in c as a result of that probability:

The difference in this equation Eq. (24), and Eq. (6), regarding gravitation, is that this is the mutual change in the photons velocity resulting from a mutual change in the oncoming photon density whereas Eq. (6) is the effect of the ambient probability density or the effect on an interloping photon. This expression, Eq.(24), provides the mutually induced gradient in c for the photons that holds the two photons in orbit in the electron. Understanding this expression is the key to understanding how the universe works

Structure of the Electron

Reducing Eq. (24), gives:

$$\frac{\Delta c}{c_0} = \left(\frac{\lambda_{PL} v_e}{r_{12}} \right)^2 \tag{25}$$

Then the minimum value of r is:

$$\frac{\Delta c}{c_0} = \left(\frac{\lambda_{PL} v_e}{r_{12}} \right)^2 \rightarrow r_{12} = \lambda_{PL} v_e \tag{26}$$

Thus At the minimum probable value of r, $\Delta c = c_0 - c = 0$, **the photons would be stopped.** This is below the possible orbiting radius for the electron, since energy and angular momentum could not be conserved. Is the radius at which the **flux density of the two pho-**

tons inside the electron increases the ambient flux density to the point that the photon mean free path becomes zero.

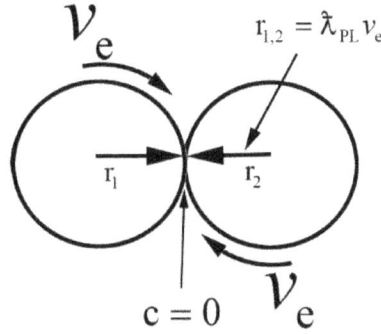

Fig. 1. The Electron Planck particles at their minimum r$_{min}$, distance would bring the relative velocity to c = 0,

The minimum orbit calculated here is related to the electron minimum orbit, but angular momentum and energy can't exist if the velocity is zero. The actual minimum occurs when the sum of the angular momentum is $\hbar/2$, and the velocity of light is $c = c_0/2$

The two orbiting photons in the electron are bosons and as such have an angular momentum of $\hbar$. The in angular momentum, p is a function of c, thus when the velocity of light is 1/2, the angular momentum is $\hbar/2$, creating a Fermion. This is also the limitation that defines the Fine Structure Constant.

Eq. (25), can be re configured as:

$$\frac{\Delta c}{c_0} = \frac{1}{2}\left[\left(\frac{\sqrt{2}\lambda_{PL}v_e}{\lambda_e}\right)\left(\frac{\lambda_e}{r}\right)\right]^2 \qquad (27)$$

The value of $\Delta c/c_0$ Is ½ so the terms in the bracket is equal to 1

Although not intuitive the group of constants in the first parenthesis is identified as the **fine structure constant** α,

$$\alpha = \frac{\sqrt{2}\,\lambda_{PL}v_e}{\lambda_e} \qquad \text{(See Eq.(32), and appendix I)} \qquad (28)$$

Eq. (27), then becomes.

$$\frac{\Delta c}{c_0} = \frac{1}{2}\left(\frac{\alpha \lambdabar_e}{r}\right)^2 \tag{29}$$

, and the minimum value of r is

$$r = \alpha \lambdabar_e = \sqrt{2}\, \lambda_{PL} v_e \tag{30}$$

This is the $\sqrt{2}$ larger than the stopping radius of the photons Eq. (26). It is the classical radius of the electron, and defines the fine structure constant, Eq. (28).

Atomic Energy Levels

With the parameters for the electron defined above it is a short step to defining the energy levels of an atomic system of two particles.

In the case of two attracting opposite charged mass particles, the electron and the positron, the same mechanisms as for the binding of the photons in the electron apply. The coincidence of the interaction of the two photons in the opposite particles provides the gradient in c that binds the two particles. The coincidence probabilities of angular momentum quantum levels correspond to eigenvalues of the wavefunction in the Schrodinger equation.

The Feynman photons generating the probability density for the two photons in the electron, is exactly equivalent to the interaction probabilities of the two mass particles. Noting that . $2v_{1,2} = v_e$, Eq. (24), becomes:

$$\frac{\Delta c}{c_0} = \frac{\lambdabar_{PL}^2}{\lambdabar_e^2}\left(\frac{2\lambdabar_e}{r_1}v_1\, \frac{2\lambdabar_e}{r_2}v_1\right) \rightarrow \frac{\lambdabar_{PL}^2 v_c^2}{\lambdabar_e^2}\left(\frac{\lambdabar_e}{r_1}\, \frac{\lambdabar_e}{r_2}\right) \tag{31}$$

The difference is that for massive particles the minimum value of the separation of the particle distances is the Compton radius, $\lambda_e{}^*$. This is the radial interaction point where the opposite going photons have coincidence and have integral but opposite angular momentum. The photon probabilities can be anywhere but at this radius is the probability of the coincidence of the photons of the two particles. Max Born, et, al, identified what is referred to here, Eq.(20), as of the photon probability density as the wavefunction amplitude, and the photon coincidence probability density, Eq.(22), as the square of the wavefunction amplitude.

The major difference between Eq.(25), for the bound photons, and Eq.(31), for two mass particles is that the first is bound bosons having only a single quantum level of $\hbar/2$ and the latter is pair of fermions can have multiple discrete values of angular momentum.

Reorganizing Eq.(31), again gives

$$\frac{\Delta c}{c_0} = \frac{1}{2}\left[\left(\frac{\sqrt{2}\,\lambda_{PL}v_e}{\lambda_e}\right)\frac{\lambda_e}{r_1}\right]\left[\left(\frac{\sqrt{2}\,\lambda_{PL}v_e}{\lambda_e}\right)\frac{\lambda_e}{r_2}\right] \tag{32}$$

*Note: D_e is the actual radius of the orbiting particle, and includes the anomalous g_A factor that defines the Electron Creation Radius, $\mathfrak{R}_0$ noted in later papers.

After some pre-anticipated factorization the terms in parenthesis are identified as the fine structure constant Eq.(28), (Appendix I), and equation E.(32), becomes:

$$\frac{\Delta c}{c} = \frac{1}{2}\left(\frac{\alpha\lambda_e}{r_1}\right)\left(\frac{\alpha\lambda_e}{r_2}\right) \tag{33}$$

This is the change in c at an observation point at the distance, r_1, and r_2, from the center of the two particles 1 and 2.

Two Particle Energy Potential

For potential energy considerations, setting the observation point r_1 for particle 1 at the point of minimum energy, $\alpha\lambda_e$ Eq.(30). The value of Δc is then the difference between the minimum value at the first particle, $c_0/2$, and the change as a function of the position of the second particle. This then provides the specific energy change per unit mass of the potential energy as a function of the distance between the particles,.

$$\frac{\Delta c}{c_0} = \frac{1}{2}\left(\frac{\alpha\hbar c}{m_e c^2 r_1}\right) = \frac{1}{2}\frac{Q^2}{m_e c^2 r} \tag{34}$$

This equation gives the total energy difference as being half the conventional electric potential energy as a function of the radial distance.

$$\Delta E = mc_0^2 - m_0 c_0 c = \frac{1}{2}\frac{Q^2}{r} \tag{35}$$

The value of Q^2 is, $\alpha c\hbar$ and allows identification with the electric potential but it is not the product of charges, and is just the compact representation of the terms

Eq.(35), provides a potential energy difference that binds the particles together expressed as an electric equivalent.

By taking the gradient of ε, in Eq.(35), shows the force due to the gradient in the relativistic energy to be, the gradient in the total energy, not the change in the electric field.

$$f = \frac{d\varepsilon}{dr} = m_e c_0\left(\frac{dc}{dr}\right) = -\frac{1}{2}\frac{Q^2}{r^2} = \frac{QE}{2} \tag{36}$$

This is half the electric field potential energy. The Lagrangian electric field equivalent of this expression is that the total energy difference is:

$$\phi(\varepsilon) = -\frac{1}{2}\frac{Q^2}{r} = -\frac{Q^2}{r} + \frac{1}{2}mv^2 \tag{37}$$

Atomic energy levels, A New Atom

Equation Eq.(33), gives the continuous probability densities of the Feynman photons produced by the two charged particles, and the relative energy differences at any distance. The electrons although composed bosons are fermions and can only have specific values of angular momentum:

$$\frac{\Delta c}{c} = \frac{1}{2}\left(\frac{\alpha}{m_e c r_1}\frac{\hbar}{}\right)\left(\frac{\alpha}{m_e c r_2}\frac{\hbar}{}\right) \tag{38}$$

This means that their Compton radii are integral values, and the angular momentum is a multiple of $\hbar$, that is: $L_1 = m_e c r_1 = n_1 \hbar$.

From an observation point at the closest approach of the two particles to each other, or at the point that their radii touch, the change in c is:

$$\frac{\Delta c}{c_0} = \frac{1}{2}\left(\frac{\alpha}{n_1 \hbar}\frac{\hbar}{}\right)\left(\frac{\alpha}{n_2 \hbar}\frac{\hbar}{}\right) \tag{39}$$

This, as in Eq.(8), is the specific energy change per particle, thus the change in the energy of the electron:

$$m_e c \Delta c = \frac{\alpha^2}{2}\frac{m_e c^2}{n_1 n_2} = \frac{R}{n_1 n_2} \tag{40}$$

This is the change in energy of the mass of the electron if the coincident point of the interaction photons is integral values of the Compton radius from the particle center.

Eq.(40), is the first energy levels of an atomic system, or the principle quantum number. The primary differential energy levels for a hydrogen type atomic system are then:

$$\Delta \varepsilon = R\left(\frac{1}{n^2} - \frac{1}{n^2}\right) \tag{41}$$

These are the primary energy levels of an atom.

The separation of the center of the particles, is a multiple of the Compton radius $(n_1 + n_2)\lambda_e$, and is not multiples of the Bohr orbital radii.

This is clearly not the Bohr atom. The orbits represent the coincident probability of opposite going photons with the same angular momentum, and are the contact point is location of the standing waves of the Schrodinger equation.

RED To go out as bad conceot

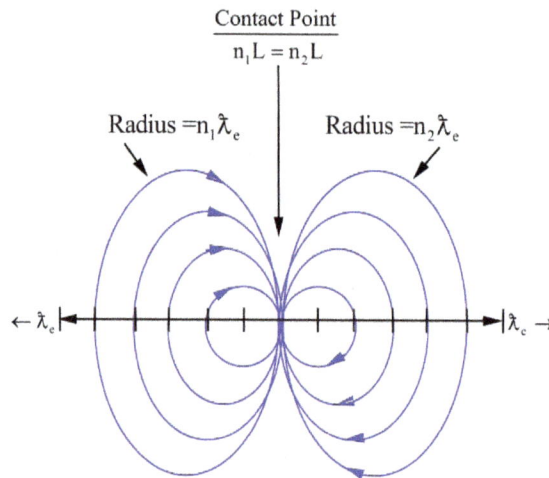

Contact Point
$n_1 L = n_2 L$

Radius $= n_1 \lambda_e$ Radius $= n_2 \lambda_e$

Fig. 2, The structure of the atom defined by Feynman photon considerstions

The contact point is the location where the angular momentum of the two particles is equal, and the Feynman photons are going in opposite directions. At this point the photons can have conjugate wavefunctions, and the energy of the mass of the electrons differs from rest energy, by the state energy level solutions of the Schrodinger equation. The location of the contact point does not have to move as the result of a state transition, but the angular momentum of the system is the sum of both the internal motion and the rotation of the particle around the center of mass, thus if the total is an integer:

$$L = m_e vr + n_1 \hbar = n\hbar \tag{42}$$

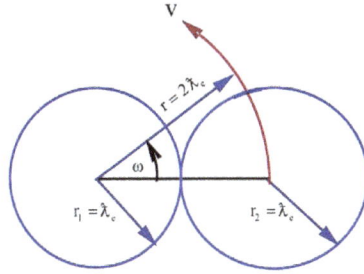

The state values can be degenerate.

$$L = m_e vr = (n_1 - n_2)\hbar \tag{43}$$

The success of the Bohr atom is seen to be due to the relation between the velocity of the electron the velocity of light, and the radius of the Bohr orbits.

$$\frac{v}{c} = \alpha \qquad r_B = \frac{\lambdabar_e}{\alpha} \qquad L_B = m_e vr_B \rightarrow L_C = m_e c \lambdabar_e = n\hbar \tag{44}$$

Positron

The Positron has a different coefficient of energy as a function of r, being R/2, rather than that of an atomic system with a heavy hydrogen center. This is the result of a change in, (Δr), for two equal mass particles, also produces a move the center of mass by half that amount $\Delta r / 2$, thus reducing the energy differential by the same factor. This is the same as the orbital energy of a reduced mass system.

This section on electric particle interaction only addressed the absolute value of an attractive charge, thus only the effect of opposite charges. Since the presented theory has no +/- signs the description of repulsive action is missing here. The repulsion of alike particles has the same but more involved causal mechanism related to spin and Lorentz properties. (See Endnote 5)

E = mc² in a Variable Speed of Light

In expressions, Eq.(9), and Eq.(37), the change in the energy $E = mc^2$ as a result of a change in c is proportional the total energy, and therefore in a conservative system includes both potential and kinetic energy.

$$\Delta\varepsilon = mc_0^2 - mc_0 c \rightarrow \quad \frac{d\varepsilon}{dr} = m_0 c_0 \frac{dc}{dr} = \phi(r) \tag{45}$$

The potential is not the electrical potential, but the change in the total energy, the sum of the total change in relativistic rest and kinetic energy in a conserved system.

$$\phi(r) = \Delta m_0 c_0^2 + \frac{1}{2}mv^2 \tag{46}$$

For a particle moving into a zone of altered c the rest mass decreases and the velocity increases, For a conservative system in which the energy is constant the decrease in rest mass energy is offset with the increase in kinetic energy, thus the relativistic energy and mass remain constant.

$$m_0 c_0 c \left(1 + \frac{1}{2}\frac{v^2}{c^2}\right) = m'c^2 \quad \text{constant} \tag{47}$$

This perspective is not the normal view of relativist mass and energy. Normally the relativistic mass does not involve a change in c, and the acquired energy is by acceleration. For a conservative system, there is no acquired or lost energy in moving in or out of a zone of an increased index of refraction, thus the total relativistic energy is constant. The kinetic and rest energy have an inverse proportionality.

Eq.(45), shows a slightly changed view of the relativistic mass. It is illustrated that the square of c^2, is actually the product of two value of c, one of the observer, c_0 and one in the observed frame, c.

Relation of Schwinger Vacuum Polarization Electric Field Density and the Photon Flux Density

Energy Density E_{CR}^2 Feynman Photon

Flux Density n_f

The purpose of his section is to show the relation between the Schwinger vacuum electric energy density, and the Feynman photon induced speed of light in the universe.

In 1951 Julian Schwinger developed the energy density associated with the Vacuum Polarization, showing the critical value of the electric energy density necessary for the commencement of Electron-Positron pair production to be the square of an intense electric field [12].

$$E_{CR} = \frac{m_e c^2}{Q \lambda_e} \qquad E_{CR}^2 = \frac{m_e c^2}{\alpha \lambda_e^{\;3}} = \frac{\alpha^2 m_e c^2}{\alpha^3 \lambda_e^{\;3}} \qquad (48)$$

In order to compare the Schwinger electric energy density to the photon flux density concept we need to establish the ratio of the electric field energy density to the Feynman photon density. (Endnote 6)

This can be found in the electric field density of the minimum energy density of an atomic system in which the energy density is defined.

The electron consists of two orbiting photons, thus the maximum photon probability density defined in Eq.(40), is equivalent to the electric energy induced by the probability of two orbiting

173

photons and occurs when the angular momentum of each photon is equal to $L = mcr = \hbar$. This turns out to be the Rydberg energy, and the minimum energy of the Bohr atom.

For one photon the corresponding electric energy density induced by a single photon is then half R or R/2.

Dividing the Schwinger Critical electric energy density, by the electric energy density created by a single orbiting photon value, yields the Schwinger equivalent photon number density.

$$n_{CR} = \frac{E_{CR}^2}{R/2} \qquad (49)$$

The number density of Feynman photons corresponding to the Schwinger Vacuum Polarization energy density calculates to be:

$$n_{CR} = 1.7874783E + 38 \qquad (50)$$

This is within 5% of the value of the Feynman photon density estimated for the universe, Eq. (19),

$$n_f = 1.6932057E + 38 \qquad\qquad n_{CR} = 1.7874783E + 38 \qquad (51)$$

The fact that these numbers are this close is probably somewhat coincidence, but being in the same ballpark addresses the credibility of the calculation.

The vacuum energy density of the universe found from the Schwinger electric energy density, (10^{123} erg/cm^3), is obviously not real, and exposes a major flaw in assigning an energy density to an electric field. The concept of charge, and electric field, is a useful mechanism for calculation the energy differentials in an atomic system, but fails outside that arena.

Conclusion

It is postulated that the Feynman (Planck) Photon probability density is the mechanism that: Sets the velocity of light; is responsible for Gravitational and Electrical interaction; creates the internal binding for the electron; and sets the energy levels in an atom.

Using the concept of, Δc, (delta c), and the probability of photon interactions instead of the concept of the electric field, removes the infinities of Quantum Electrodynamics that makes the analysis of the internal dynamics of elementary particles impossible. The concept of continuous fields containing energy, create unsolvable issues that stops the progress of understanding elementary particle structure.

It has not been the purpose to explore all the fine and hyper-fine structure of the interactions, or even the implications regarding other particles. The purpose is to illustrate the basic mechanisms of a novel theory of photon and particle interaction that can be useful in developing the structure of more complex particles.

References:

More extensive lists of on referenced to items, are found in the reference 1-6 by Author. Some of the papers 1, 2 are a little dated and need revisions.

1. DT Froedge, The Dirac Equation and the Two Photon Model of the Electron revised, April 2021, DOI:10.13140/RG.2.2.19095.70564, https://www.researchgate.net/publication/350922403

2. DT Froedge, The Connection between Electric Charge, Gravitation, and the Feynman Sum over All Histories View of Quantum Electrodynamics, April 2020 Conference: APS April. 18-21, 2020 Washington DC.
https://absuploads.aps.org/presentation.cfm?pid=18355 https://www.researchgate.net/publication/341310206

3. DT Froedge, A Quantum Theory Conjecture on the Origin of Gravitational and Electric Particle Interaction, December 2019, DOI: 10.13140/RG.2.2.29097.54884
https://www.researchgate.net/publication/337826826

4. DT Froedge, The Electron as a Composition of Two Vacuum Polarization Confined Photons, April 2021, DOI:

10.13140/RG.2.2.18971.18722 , https://www.researchgate.net/publication/350740864

5. DT Froedge, The Gravitational Constant to Eleven Significant Digits,
March 2020, DOI: 10.13140/RG.2.2.32159.38564
https://www.researchgate.net/publication/339943651

6. DT Froedge, The The Fine Structure Constant from the Feynman Path Integrals, March 2021, DOI:10.13140/RG.2.2.12979.55846,
https://www.researchgate.net/publication/350188862

7. Dirac, P. A. M. (1934). "Discussion of the infinite distribution of electrons in the theory of the positron". Cambridge Phil. Soc. 30 (2): 150–163. Bibcode:1934PCPS...30..150D. doi:10.1017/S030500410001656X.

8. W. Heisenberg, (1934), "Bemerkungen zur Diracschen Theorie des Positrons", Zeitschrift für Physik (in German). 90 (3–4): 209–231. Bibcode: 1934ZPhy.90..209H, doi:10.1007/BF01333516. ISSN 0044-3328. S2CID 186232913.

9. R. Serber, (1935). "Linear Modifications in the Maxwell Field Equations". Phys. Rev. 48 (1): 49–54. Bibcode:1935PhRv...48...49S. doi:10.1103/PhysRev.48.49.

10. Sauter, "Über das Verhalten eines Elektrons im homogenen elektrischen Feld nach der relativistischen Theorie Diracs", Zeitschrift für Physik, 82 (1931) pp. 742–764.doi:10.1007/BF01339461 (https://dx.doi.org/10.1007%2FBF01339461)

11. E. Uehling, A. (1935). "Polarization Effects in the Positron Theory". Phys. Rev. 48 (1): 55–63. Bibcode:1935PhRv...48...55U. doi:10.1103/PhysRev.48.55

12. J Schwinger, On Gauge Invariance and vacuum polarization , Physical Review volume e82, 1951

13. D. Valev, Estimations of total mass and energy of the universe, arXiv:1004.1035v1 [physics.gen-ph] 7 Apr 2010

14. Schwinger, J. (1949). "II. Vacuum polarization and self-energy". Physical Review. Quantum Electrodynamics. 75 (4): 651–679.

15. J Solà, Cosmological constant and vacuum energy: old and new idea, 2013 J. Phys.: Conf. Ser.453 012015, https://iopscience.iop.org/article/10.1088/1742-6596/453/1/012015/pdf

16. I. Shapiro; Gordon H. Pettengill; Michael E. Ash; Melvin L. Stone; et al. (1968). "Fourth Test of General Relativity: Preliminary Results". Physical Review Letters. 20 (22): 1265–1269. Bibcode:1968PhRvL..20.1265S. doi:10.1103/PhysRevLett.20.1265.

15. R. Blandford, K. S. Thorne, Applications of Classical Physics, (in preparation, 2004), Chapter 26 http://pmaweb.caltech.edu/Courses/ph136/yr2012/1227.1.K.pdf

16. R. Oldershaw, Towards A Resolution Of The Vacuum Energy Density Crisis, Journal of Cosmology, vol. 17, November 2011, pgs, 7359-7375, https://arxiv.org/pdf/0901.3381.pdf

17. G. Gabrielse et al,, New Determination of the Fine Structure Constant from the Electron g Value and QED, Phys. Rev. Lett. 97, 030802 (2006) http://hussle.harvard.edu/~gabrielse/gabrielse/papers/2006/NewFineStructureConstant.pdf

18 G. Gabrielse et al Cavity Control of a Single-Electron Quantum Cyclotron Measuring the Electron Magnetic Momen, arXiv:1009.4831v1 [physics.atom-ph] 24 Sep 2010

19. T. Quinn et.al, The BIPM measurements of the Newtonian constant of gravitation, G https://royalsocietypublishing.org/doi/pdf/10.1098/rsta.2014.0032

20. C. Merkatas et. al.,Consensus Value for the Newtonian Constant of Gravitation

Xiv:1905.09551v1 [physics.data-an] 23 May 2019

21. J. Evans, M. Rosenquist, F=ma Optics, American Journal of Physics 54, 876 (1986); https://doi.org/10.1119/1.14861

22. M. Urban, F. Couchot, X. Sarazin, A. Djannati-Atai, The quantum vacuum as the origin of the speed of light, EPJ manuscript, arXiv:1302.6165v1 [physics.gen-ph] 2013

Appendix I

Experimental Accuracy of the Calculated Fine Structure Constant

The value of α identified in Eq.(28), is a constant arising from the structure of the electron, It the value in terms of other fundamental constants, all known to at least 11 significant digits without any arbitrary factors except the least accurate being the gravitational constant. The gravitational constant's experimental accuracy is limited to about 3 digits with considerable experimental scatter beyond that, thus there is a limit to the extent the accuracy can be tested.

The value of the Planck length, $\lambda_{PL} = \sqrt{\mu\lambda}$, contains the gravitational constant, big G, therefore solving Eq.(28), and including the anomalous gyromagnetic adjustment for λ_e wavelength (Appendix II) for the gravitational constant gives:

$$G = \frac{\alpha^2 2\pi^2 c \left(\lambda_e g_A\right)^4}{\hbar} \tag{52}$$

If the theory is correct the value of G can be calculated to about 11 places, and compared with the experimental results. The experimental values that are in best agreement are those of BIPM and the calculated value is within the error bars of all their measurements.

Reference values of Constants and sources used in calculation (CGS units)

$$c = 2.9979245800E+10*$$

$$g_A = g_e / 2 = 1.00115965218 \dagger$$

$$\hbar = 1.054571817646E\text{-}27 *$$

$$\lambda_e = \hbar / m_e c = 3.8615926794E\text{-}11$$

$$\alpha = 1/137.035999084 \dagger$$

$$m_e = 9.1093837015(28)E\text{-}28 \oplus$$

* Definition† Gabrielse et. al. [17][18]. $\oplus$ 2018 Codata Recommended Values

Comparing calculated value of G with BIPM* and Codata consensus values (MKS)

Calculated value	G = 6.6755052700E-11
BIPM 01-32-2001	G = 6.67559(27) x 10-11
BIPM weighted mean 2014	G = 6.67554(16) x 10-11
BIPM Sep, 2015	G = 6.67545(18)) x 10-11
Codata Consensus Value	G = 6.67430(15) x 10-11

*BIPM- Bureau of International Weights and Measurements, [19].

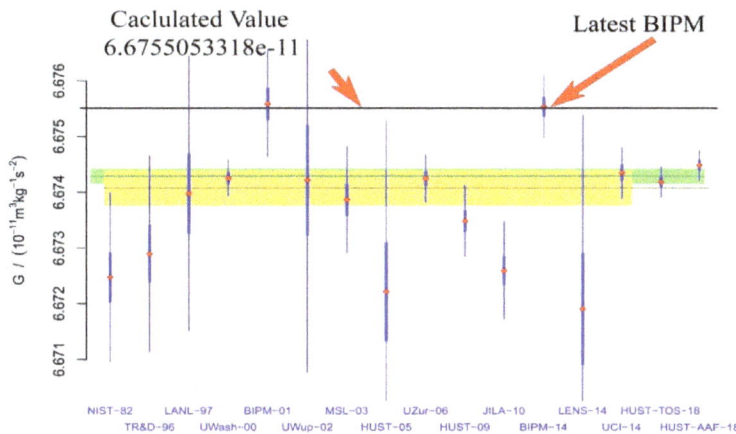

Experimental values used for statistical calculation of Codata consensus value of G, [20]

The calculated value is within 2 ten-thousands of the Codata consensus value, and within the experimental error value of the BIPM measurements.

Appendix II

Compton Radius Adjustment

The Electron developed in [4] is a composition of two orbiting photons bound by their mutually generated change in the radial index of refraction. The most probable orbit is at the Compton radius and the energy of each photon is $\frac{1}{2} m_e c_0^2$, and m_e is the rest mass of the electron.

In accurate calculations involving the Compton radius or the frequency, the Compton radius, λ_e calculated from the electron rest mass needs a correction. Due to the delay in completing the orbits by the Feynman external path loops it is slightly larger than that calculated from the rest mass. The delay is responsible for creating the Anomalous Gyromagnetic Ratio, and creating the extremal probability density of the Feynman photons, and for accuracy must be included in, $\lambda_e = \hbar / m_e c$, thus the radius is:

$$\lambda_e \rightarrow \lambda_e g_A = \lambda_e \times 1.00115965218073 \tag{53}$$

Endnotes:

1. The energy in the electric field is arbitrarily assigned based on the electric potential energy of a charge at the classical electron radius, $\alpha \lambda_e$. The potential being equal to the rest mass energy of the electron at that radius. It is then integrated over all space and then assigned an electric field energy density. The arbitrary assignment of an energy density to a continuously differentiable electric field creates an impediment to understanding the mechanics of quantum physics. The concept of electric charge and electric fields only has merit down to the Compton radius and should not be ascribed to having energy density. As Einstein pointed out there are clearly photons in a dynamic electric field, but the same is not true of an infinitely divisible static field, the assignment of an energy density is quite arbitrary, and the idea of

charge and electric field energy is a useful and intuitive concept, but it has fatal theoretical flaws.

2. The velocity change in the speed of light by gravitation Eq. (7), has been known for some time, but what seems to have been overlooked is that a change in c causes a change in the energy of mass particles via. $E = mc^2$ The change in c thus is shown as the source of gravitational potential energy as well as electrical potential.

   Eq. (7), is the interaction effect of one mass on the other, and is not reciprocal. For large differences in mass this is correct, however for large close masses interactive effects may have to be considered. The total interactive expression needs the product of the interaction as is developed for the electron Eq. (23).

3. *Speed of light in an empty universe*

   It was noted that the speed of light is proportional the number of Planck particle hit to misses per second as the photon moves at c, or:

$$\frac{c}{c_0} = \frac{n \times \text{Area of PlanckPaeticle}}{1 \text{ cm}^2}$$

From Eq. (19), the number of protons photon per square cm per second n_f times the Planck particle gives the ratio to be $3.9881E-26$. With this, the current value, and the vacuum value of c can be found noting that the current value of c and the change in c is the result of the photon number

$$\frac{c_{00} - c_0}{c_{00}} = \left(1 - \frac{\lambda_{PL}^2 n_f}{1 \text{cm}^2}\right) = (1 - 3.988E - 26)$$

Where, c_0 is the current velocity of light, and c_{00} is the value in an empty universe.

$$c_{00} = 7.5171e35 \text{ cm/sec}$$

This is about 60 million times across the universe in a second. The current value is:

$$c_0 = c_{00} 3.9881388E - 26 = 2.9979245800E\ 10 \text{ cm/sec}$$

The value of c is proportional to the mass and the radius of the universe, both related to c, thus the value of c, may be quite complicated.

181

The velocity of light is dependent on the cross section of the opposing Planck particles which are independent of the frequency, thus the lack of frequency dependence on the speed of light in vacuo is apparent.

4. Previous papers of the author have erred in putting in the Compton frequency of the electron in this expression, leading to an error in the potential. The frequency should be the Compton frequency of the two orbiting photons in the electron which is ½ of the electron frequency. The error leads to the derived potential being equal to the electric potential, which is not the total energy due to the change in c. Putting the proper frequency of the photons yields the potential being that of the total energy change due to Δc.

5. This paper has only addressed the concept of attractive charges or the absolute magnitude of the interactions. Since the concept of presented here does not have a +/-, Q_1, Q_2, the repulsion of alike particles does not have a similar mechanism. The mechanism equivalent to charge is inherent in the nature of the Lorentz interaction of photons and the spin associated with the Electron and the Positron. The photon interactions presented here are for photons moving in opposite directions creating attractive forces, whereas photon encounters moving in the same direction do not interact. This can be better presented in terms of photon wavefunctions and Geometrical Algebra first defined by Dirac. (See "The Dirac Equation and the two Photon Model of the Electron. [1].).The charge sign alternative mechanism will be presented in a subsequent paper.

6. There is a correspondence between the gradient in the index of refraction and the force holding a photon in orbit [21], and for comparison, it can be notices that the Schwinger critical force on charge is equal to the centrifugal force on a photon rotating at the Compton radius:

$$f = QE_{CR} = \frac{pc}{r} = \frac{m_e c^2}{\lambdabar_e} \qquad (54)$$

Calculated Value of Gravitational Constant from Fine Structure Constant and Electron Mass

D.T. Froedge

V060219

In the Charged Particle Interaction Section of "A Quantum Theory Conjecture on the Origin of Gravitational and Electric Particle Interaction"[1], there is a relational value between α and other fundamental constants including The Gravitational Constant and the Electron mass, that can be useful in testing the theory.

$$\alpha^2 = \frac{2\lambdabar_{PL}^2}{\lambdabar_e^2} v_e^2 \tag{1}$$

This is the value of the fine structure constant, Alpha(α), in terms of fundamental constants that are known to a high degree of precision by calculations from QFT considerations. The $\lambdabar_e$ is the Compton radius of the electron, and v is the Compton frequency $v_e = c/(2\pi\lambdabar_c)$. The Planck length is:

$$\lambdabar_{PL}^2 = \frac{G\hbar}{c^3} \tag{2}$$

The value of α^2 is nominally:

$$\alpha^2 = (1/137)^2 \qquad (3)$$

or more accurately from the work of Gabrielse et.al.[2],

$$\alpha = 0.00729735253594 \qquad (4)$$

Whether this theory has merit or not, hinges to a degree on the accuracy of the prediction of this relation. This value of α (Eq.(1)), is expressed in fundamental constants known to a high accuracy (~12 significant digits) and there are no

"fudge" factors, The major uncertainty in the calculationis the value of the gravitational constant G, which is experimentally determined to only about 5 significant digits, is the major uncertainty, the gravitational constant..

A more explicit for Eq.(1), is:

$$\alpha^2 = \frac{G\hbar}{c^3} \frac{2}{\lambda_e^2} \frac{c^2}{2^2 \pi^2 \lambda_e^2} == \frac{G\hbar}{2\pi^2 c \lambda_e^4} \qquad (5)$$

The model for the electron that generates the relation for α is of two photons revolving round the center of momentum, and just as an in the case of an electron revolving around the proton, the quantum loops of the Feynman paths of the orbiting photons must be taken into account. The mass calculated Compton wavelength is the first loop, and the effect on the wavelength of all the other loops must be taken into account, this is done by multiplying the Compton radius λ_e, by the anomalous gyromagnetic ratio $g_e/2$

$$g_e/2 = 1.0011596522 \qquad (6)$$

To assess the predicted theoretical value, the expression Eq.(5), can be solved for the gravitational constant, and compared with current experimental values.

Solving for the electron mass in terms of fundamental constants gives:

$$m_e = \frac{\hbar g_e}{c} \left(\frac{2\pi^2 c \alpha^2}{G\hbar} \right)^{1/4} \qquad (7)$$

Solving, for the gravitational constant gives:

$$G = \frac{\alpha^2 2\pi^2 c (\lambda_e g_e)^4}{\hbar}$$

(8)

Calculating G from Eq. (8), to 12 significant digits gives:

$$G = 6.67586727955 \times 10\text{-}08 \text{ cm}^3\text{gm}^{-1}\text{s}^{-2}$$

This is slightly higher (0.02%) than the current Codata consensus recommended value of $6.67430(15) \times 10^{-11} \text{ m}^3\text{kg}^{-1}\text{s}^{-2}$, but it is not outside the scatter of current measurements, and is exactly the value determined by the Cavendish balance measurements of the International Bureau of Weights and Measures, BIPM value by T. Quinn et.al,[3], published in 2015.

$$G = \underline{6.67586}(36) \times 10\text{-}11 \text{ m}^3\text{kg}^{-1}\text{s}^{-2}$$

The calculated value is in agreement with this value to the limits of its experimental accuracy of this measurement.

Conclusion

Accept for some numerology associations, the equation above, Eq. (1), is the only known physical relation between the fine structure constant, the Gravitational constant, and the mass of the electron. The scatter in the experimental value of G is the only uncertainty.

Reference values:

Constants used in calculation in CGS units.

$c =$	2.9979245800E+10	$g_F = g_e / 2 =$	1.00115965218
$\hbar =$	1.0545918473E-27	$\lambda_e = \hbar / (m_e c) =$	3.8616633678E-11
$\alpha =$	1/137.03599971	$m_e =$	9.109389966E-28

Summary:

Calculated value of G	$G = 6.67586727955 \times 10\text{-}11 \text{ m}^3\text{kg}^{-1}\text{s}^{-2}$
BIPM Cavendish balance value	$G = 6.67586(36) \times 10\text{-}11 \text{ m}^3\text{kg}^{-1}\text{s}^{-2}$
Codata Consensus Value	$G = 6.67430(15) \times 10\text{-}11 \text{ m}^3\text{kg}^{-1}\text{s}^{-2}$

Excerpt from BPIM report: *https://royalsocietypublishing.org/doi/pdf/10.1098/rsta.2014.0032*

PHILOSOPHICAL TRANSACTIONS OF THE ROYAL SOCIETY A

rsta.royalsocietypublishing.org

The BIPM measurements of the Newtonian constant of gravitation, *G*

Terry Quinn[1,+], Clive Speake[2], Harold Parks[1,+] and Richard Davis[1,§]

12. A value for Newton's constant of gravitation

The peak-to-peak servo torque, τ_s, obtained as an unweighted mean of 10 data runs was $3.148869(94) \times 10^{-8}$ N m and using equations (9.2*a*) and (11.3) we can write

$$G_s = \frac{\tau_s}{\Gamma_s} = 6.67515(41) \times 10^{-11}\,\mathrm{m^3\,kg^{-1}\,s^{-2}}\ (61\,\mathrm{ppm}). \tag{12.1a}$$

The unweighted mean of the 10 data runs giving a value of the peak-to-peak deflection angle of 0.1529322(29) mrad using equations (9.2*b*), (8.2) and (11.8) we can write

$$G_c = \frac{\tau_c}{\Gamma_c} = 6.67586(36) \times 10^{-11}\,\mathrm{m^3\,kg^{-1}\,s^{-2}}\ (54\,\mathrm{ppm}). \tag{12.1b}$$

We have used the values for the uncertainties in the experimental measurements given in §11.

Experimental Data

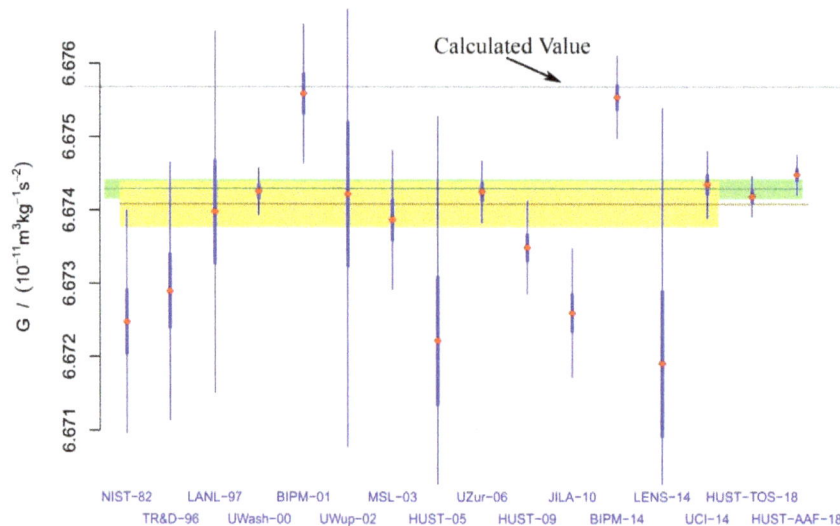

Experimental values used for statistical calculation of Codata consensus value of G [4]

Reference:

1. D.T. Froedge, A Quantum Theory Conjecture on the Origin of Gravitational and Electric Particle Interaction, http://www. arxdtf.org/css/QFTGravitationalElectric.pdf DOI: 10.13140/ RG.2.2.29097.54884

2. G. Gabrielse et al,, New Determination of the Fine Structure Constant from the Electron g Value and QED, Phys. Rev. Lett. 97, 030802 (2006)
http://hussle.harvard.edu/~gabrielse/gabrielse/papers/2006/ NewFineStructureConstant.pdf

3. T. Quinn et.al, The BIPM measurements of the Newtonian constant of gravitation, G
https://royalsocietypublishing.org/doi/pdf/10.1098/rsta.2014.0032

4. C. Merkatas et. al.,Consensus Value for the Newtonian Constant of Gravitation
Xiv:1905.09551v1 [physics.data-an] 23 May 2019

The Gravitational Constant to Eleven Significant Digits

D.T. Froedge

V031620

The interaction of the Feynman path photons resulting from the internal particle paths defines both gravitational and electrical forces [1],[2],[3]. As a result, the value of the fine structure constant is found expressed in terms of

fundamental constants as:
$$\alpha^2 = \frac{2\lambdabar_{PL}^2}{\lambdabar_e^2} v_e^2$$

$$(1)$$

The constants are the Planck radius, $\lambdabar_{PL} = \sqrt{\mu\lambdabar} = \sqrt{G\hbar/c^3}$, the Compton radius $\lambdabar_e$, the anomalous gyromagnetic ratio $g_A = g_e/2$, and the electron Compton frequency $v_e \to m_e c^2/2\pi\hbar$. The Feynman photons of two interacting particles represent the first order quantum loop, and thus the wavelength and the reciprocal of the frequency of the electron must be corrected by the higher order perturbations included in the anomalous gyromagnetic ratio, $g_A = g_e/2$, and $v_e \to v_e/g_A$. The values of the constants are included in Table [1].

All of the constants in the expression are known to at least eleven significant digits, except the gravitational constant. The maximum number of digits forthe gravitational constant known with experimental certainty is about three.

By solving for the gravitational constant (G) in Eq.(1), the expression becomes:

$$G = \frac{\alpha^2 2\pi^2 c \left(\lambda_e g_e\right)^4}{\hbar} \tag{2}$$

From these values the gravitational constant can be calculated to an accuracy of about eleven significant digits:

$$\underline{G = 6}.6755053318 \times 10\text{-}8$$

This value is slightly higher (0.02%) than the current Codata consensus recommended value of 6.67430(15) x10- 11m3kg-1s-2, but it is not outside the scatter of measurements used in forming that consensus. It is within the error bars of all the experimental measurements conducted by the International Bureau of Weights and Measures, (BIPM), published since 2000, [4][5][6[[7]. Comparison with those values is presented below, Table [2].

The merit of this theory requires this relation to be correct. If it is wrong the theory is flawed.

Table 1
Constants and sources used in calculation, CGS units.

$c = 2.9979245800E+10$*	$g_A = g_e/2\ = 1.00115965218$ †
$h = 1.054571817646E\text{-}27$ *	$\lambda_e = \hbar / m_e c = 3.8616633678E\text{-}11$
$a = 1/137.035999084$ †	$m_e\ \ = 9.1093837015(28)E\text{-}28$ ⊕

* Definition † Gabrielse et. al. [8][9]. ⊕ 2018 Codata Recommended Values

Table 2

Comparing calculated value of G with BIPM and Codata consensus values

Change in h-bar should be	6.6759940754
Calculated value	G = 6.6755053318 x 10-11
BIPM weighted mean 2014	G = 6.67554(16) x 10-11
BIPM Sep, 2015	G = 6.67545(18)) x 10-11
BIPM 01-32-2001	G = 6.67559(27) x 10-11
Codata Consensus Value	G = 6.67430(15) x 10-11

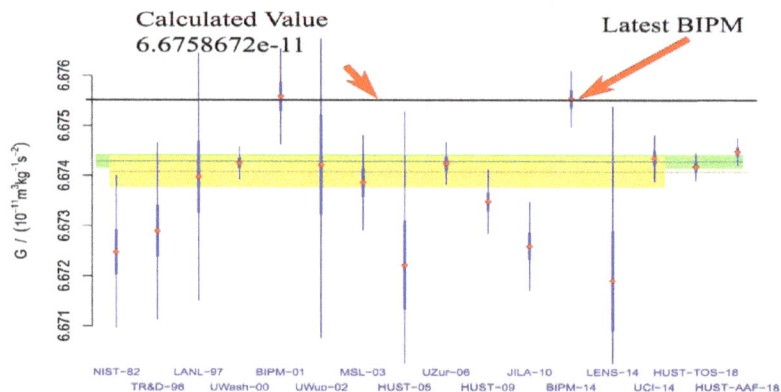

1. DT. Froedge, The Electron as a Composition of Two Vacuum Polarization Confined Photons Revised, February 2020, DOI: 10.13140/RG.2.2.31942.22085, https://www.researchgate.net/publication/339512823

2. DT. Froedge, A Quantum Theory Conjecture on the Origin of Gravitational and Electric Particle Interaction, December 2019, DOI: 10.13140/RG.2.2.29097.54884, https://www.researchgate.net/publication/337826826

3. DT. Froedge, Quantum Field Origin of Gravitation, APS, April, 2019; Denver
http://meetings.aps.org/Meeting/APR19/Session/H11.6

4. DT. Froedge, The Gravitational Constant to Eleven Significant Digits, March 2020, https://www.researchgate.net/publication/339943651

A Quantum Theory Conjecture on the Origin of Gravitational and Electric Particle Interaction

D.T. Froedge

V090819

Abstract

Gravitation defined in curved space has never been found to be compatible with electromagnetic theory or quantum mechanics. This paper presents a theory of gravitation and electric interaction constructed within the locally conserved concepts of QFT.

It is proposed that a photon moving through probability density amplitude of approaching Feynman photons experiences an alteration in the index of refraction. This alteration of photon dynamics can be shown to be the causation of gravitation, electric charge, the structure of the electron, and the velocity of light.

In the 1950's Feynman developed what is referred to as the path integral, or sum over all histories approach to QFT this asserts that the action path taken by a particle or photon is a composition of an infinite number of action paths, the classical path being the most probable. In essence these paths represent the amplitude probability distribution of the particle as it moves through space, and as such represent a probability of the particle being on those paths.

Order of Presentation

Introduction

Paul Davies in his introduction to *Six Easy Pieces* by Richard P. Feynman said:

"You could not imagine the sum-over-histories picture being true for a part of nature and untrue for another part. You could not imagine it being true for electrons and untrue for gravity"[2]

If gravitation is a gradient in c as discussed by the author in other papers [3], then there must be a mechanism for inducing a change induced by a locally confined energy. This paper discusses how this is possible.

Feynman proposed that for a photon, or any particle, going from one point to another, there is a probability on arrival that the particle has traveled every possible path [2], and by very accurate measurements of quantum effects there is every reason to believe that this is true. It is not unreasonable to presume that the interaction of these photons with passing photons.

The probability of a photon is asserted to be a density amplitude and devoid of energy, but the interaction of the flow density amplitudes, of oncoming photons exchange creates the effect of particle interaction. By Lorentz consideration, photons having a negative velocity vector dot product do not interact.

Although the Dirac and Schrodinger QM waves associated with a particle are have long been accepted to be a Born probability distribution, the waves associated with a photon have generally been asserted to be electromagnetic. This is not necessary and in fact is counterproductive to understanding quantum dynamics and particle-particle interactions. The ascribing the discrete energy of a photon to be a continuous electromagnetic field is an artifact that leads to violations of special relativity, and in the case of the electron structure leads to unacceptable infinities.

Postulates

The following are the basic postulates that define the interaction of particles in the universe, and how they interact. The proposed postulates collectively yield known physically results, but are not mathematically rigorous.

The photon is presumed to be a Planck size particle with radius $\lambda_{PL}^2 = G\hbar/c^3 = \mu\lambda$. The wavefunction is defined as the future probability of the location with a cross section equal to the square of the Compton radius λ_C. The wave, generally thought of as electromagnetic, is asserted to be the flow of the amplitude probability of the Planck particle's future location and is devoid of energy content. On measurement or absorption by the Planck particle in an atom, the future probable location instantly vanishes everywhere, there being no energy content, there is no violation of Special relativity. In Quantum language, this is the collapse of the wavefunction. It is presumed to collapse each time there is a particle-particle interaction.

Wave Function Considerations

The change in the structure of QM By the proposed structure of the photon can be shown as it relates to the normalization condition of the wavefunction.

$$\int_0^\infty y * y = 1$$

For the photon this presumes that the expectation value of the energy of the photon would be:

$$E = \frac{\langle p \rangle}{c} = \int_0^\infty e^{-ikx+i\omega t} \left(\frac{\hbar}{i} \frac{\partial}{\partial x} \right) e^{ikx-i\omega t} \, d\tau$$

$$\langle E \rangle = \hbar \omega$$

The postulate presented here on the nature of the photon is that the wavefunction is only an amplitude of the of the probability flow.

$$\psi_1 = e^{-ikx+i\omega t}$$

The square of this is still imaginary and does not represent an energy density or energy flow. Our presumption is that the conjugate is an actual opposite going photon carrying a probability amplitude coincident, and such that it is a conjugate of the first.

$$\psi_1 = e^{-ikx+i\omega t} \qquad \psi_1 = e^{+ikx-i\omega t}$$

The presumption of QM has been that the product of the wave-functions times its own conjugate is the probability density of that wavefunction. The presumption here is that it is only the product of wavefunctions of separate photons that have a probability density. That is a probability density and transfer of momentum or energy only takes place when a second photon is going in the opposite direction with an opposite phase

$$\int_0^\infty y_1 y_2 \, d\tau = 1$$

Under normal conditions this is not unity for two random photons, their phases are not conjugate, and they are not in the same location. The square of a photon's wavefunction is not energy.

The normalization condition can occur when an in phase stream of photons is incident on a higher index of refraction target. The reflection reverses the phase of the first photon, and when they are coincident.

$$y_1 = y_2{}^*$$

At this point the normalization condition is satisfied and there exists an energy density that the reflecting target can absorb.

This is the condition for defining the location of two photons arriving at a target screen for the double slit, and the coincident probability at the detector, for Bell's inequality.

1

The change in the speed of light, or index of refraction, on passing through the Compton volume of another photon per unit length is proportional to the probability of a collision with the Planck core of that photon in that volume of space.

$$\frac{\Delta c}{c_0} = \frac{\lambdabar_{PL}^2}{\lambdabar^2} \qquad (1.1)$$

This is the same as the probability of hit or miss of the Planck particle in the flow number density equivalent to the photon.

$$P_{\gamma\gamma} = \frac{\Delta c}{c} = \frac{\lambdabar_{PL}^2}{\lambdabar_P^2} = \frac{\sigma_{PL}}{\sigma_c} \qquad (1.2)$$

$$\sigma_{PL} = \lambdabar_{PL}^2 = \mu\lambdabar = \frac{G\hbar}{c^3} \qquad (1.3)$$

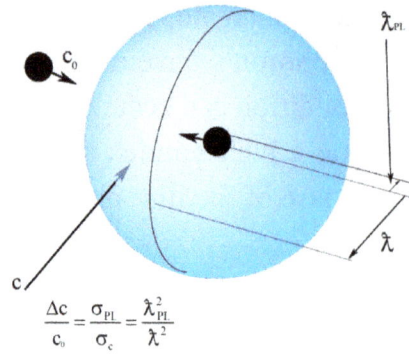

photon delay passing through flow amplitude density of another photon.

$$\frac{\Delta c}{c_0} = \frac{\sigma_{PL}}{\sigma_c} = \frac{\lambdabar_{PL}^2}{\lambdabar^2}$$

Fig.1

2

The time average probability density ratio of Feynman Photon at a point at a distance r to the probability of being at the center of mass particle is:

$$P = \frac{2\lambdabar_C}{r} \qquad (1.4)$$

(See Appendix I for discussion)

3

Multiplying Postulate #1 times postulate #2 gives the change in the speed of light as a function of the distance from the particle as the result of the density of Feynman photons at that position.

$$\frac{\Delta c}{c_0} = \frac{\lambdabar_{PL}^2}{\lambdabar_P^2} \frac{2\lambdabar}{r} \qquad (1.5)$$

This postulate is the fundamental relationship that provides the mechanisms for both Gravitational and Electric interactions and the structure of the electron. For gravitation the expression reduces to the well-known value.

198

$$\frac{\Delta c}{c_0} == \frac{2\mu}{r}$$ (1.6)

5

A particle may have many internal constituents, but the average probability of location will be the same that defined in Eq.. All massive hadron particles are considered as the ½ spin quantum number entangled confinement of a number of light speed particles that can be summed into a single rotating action path of two photons with density equivalent to the combined mass energy of the particle.

6

The photon is a Planck particle with a probability of location within the Compton radius. The wavefunction associated with the electromagnetic field is the amplitude flow probability of the particle, and is devoid of energy content. There is no continuous energy field associated with the photon.

7

Variable Speed of Light: Constants and Dependences

In a volume of space with a locally altered velocity of light such as induced by gravitation, the energy and frequency of a photon is considered to be a Lorentz universal constant and equal to $\hbar\omega$. The other fundamental relations for particles and photons must be therefore functions of c:

$$\varepsilon = \hbar\omega = \hbar\frac{c}{\lambda} = mc^2 = pc \qquad p = \frac{\varepsilon}{c} \qquad m = \frac{\varepsilon}{c^2} \qquad \lambda_c = \frac{\hbar}{mc} = \frac{\hbar c}{\varepsilon}$$

$$\mu_m = \frac{Gmc^2}{c^4} = \frac{\lambda_{PL}^2\,\omega_m}{c} \qquad \lambda_{PL}^2 = \mu\lambda \qquad \hbar = mc\lambda_c$$ (1.7)

The energy & angular momentum are invariants with respect to c, but the rest shown here are dependent. The mass of the photon is a variable of c and defined as $m = \varepsilon / c^2$, is inversely proportional to the square of the velocity of light and the momentum p is inversely proportional to c. The particle Compton wavelength is directly proportional to c, and gravitational constant, μ, is inversely proportional to c thus the product square of the Planck radius, $\lambda_{PL}^2 = \mu\lambda$, is independent of c.

The Compton wavelength λ_c is proportional to the value of c, and thus the Compton size of an electron $\lambda_e = \hbar c / mc^2 = c / \omega_e$ is determined by the local speed of light.

8

The electron is a composite of two photons bound in circular motion by the gradient in c produced by the first postulate. The circular orbit is the result of the gradient in generated by the probable interaction. The composite angular momentum of the two photons is ½ $\hbar$

(See Appendix III for summary)

The photons in the core of the electron are radially polarized such that their transverse flow probability amplitude is polarized along the radial direction keeping the photon on a most probable circular path. The Feynman photons generated by an electron or positron also have a radial polarization of the transverse flow component.

Feynman Photons

From Feynman's Path Integral formulation or, sum over all histories, of QFT, the action path of a particle from one spacetime point to another is the sum of the action paths over all possibilities, [2]. For a repetitive action path that is from one point to another and back, there is continuously regenerated set of paths resulting in the particle having a time average probability of being at a distance from the most probable path.

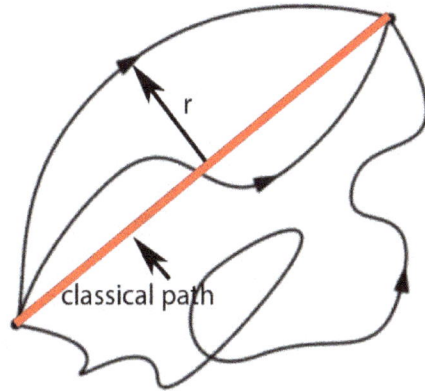

classical path

r

As a photon moves from one point to another there is the probability of being at a distance r from the classical path.

Fig.2

Since the developed of the path integral approach, it has been apparent that particles traveling on action paths have some probability of existence exterior to the classical path [4], [5], [6], [7], [8], [9] . When photons are confined such as between two reflectors or are in an orbit of mutually generated gradient in the index of refraction such as in an electron, [10],[11]. Feynman's argument requires there are multiple action paths repeating at the frequency of the cycling of the action paths throughout the surrounding space, and inducing the probability of existence of "Feynman" photons in the space surrounding the "classical" paths.

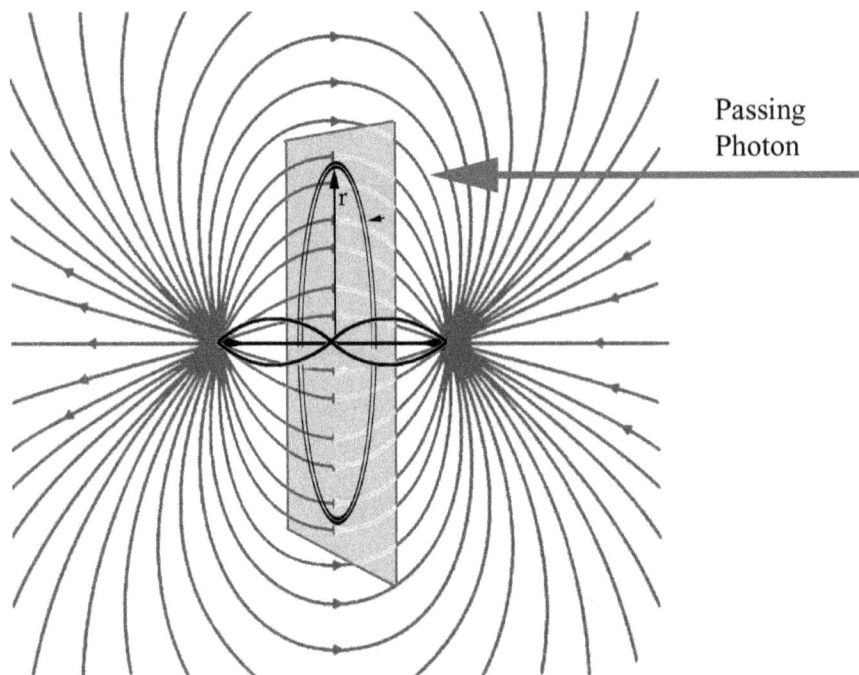

Some of the possible Feynman paths for a photon oscilating between two points.

Passing Photon

Fig.3

For an entangled pair of photons, such as exist in a particle, the exchange of momentum by external Feynman photons alters the velocity of the center of momentum, and thus exchanges both energy and momentum for the entangled pair. The probability density of the Feynman photons does have physical consequences.

The knowledge of the probabilities of discrete photons existing off the classical paths has existed more than half a century yet the exploitation of the interaction of particles has defaulted to a continuous field theory, (QFT) [11],[11a]. The continuous field is replete with infinities and is not exactly equivalent to the discrete "path integral", or "sum over all histories" approach.

As developed from work of others in and discussed in [12][13}, and in Appendix I, the time average ratio of the probability density of a Feynman photon being at the distance r from the particle, to the probability of within the Compton radius of the classical path is postulated to be:

$$P = \frac{2\lambdabar_c}{r} \qquad (1.8)$$

This is the time average relative probability density of the particle being at a point in space to the center of the action path, or the particle probable density at that point.

In the authors paper "Photon-Photon Vacuum Polarization Composite Electron Model" a model was presented showing the electron as two photons bound by an index of refraction generated by the nonlinear vacuum polarization [13]. (Summary in Appendix III)

As the photons revolve the Feynman action paths exist throughout space and repeat at each cycle of the photon circuit around the center of momentum. There is then generated a time average probability density of the photon existing at a point in space as stated in Eq. (1.8), and shown in fig. 4.

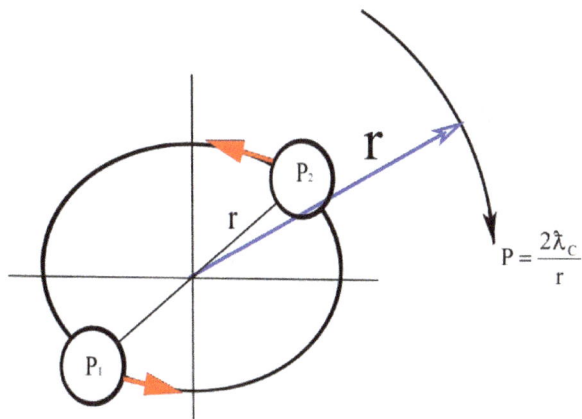

$$P = \frac{2\lambdabar_C}{r}$$

Shows the time average probability density for a Feynman photon on an action path

Fig. 4

From Eq. (1.5), and Eq. (1.6), the presumption is made that this probability density is a general property of matter and applies to all mass particles.

Photons orbiting inside mass particles are not necessarily axially aligned with other particles and do not have a common rotation axis thus this expression is the time average density of all random directed photons and constitutes the probability density of photons generated by a mass.

Although a composite particle can consists of a number of discrete internal constituents, the time average probability of the particle being at a point is the same as that from the action path of two orbiting particles with the sum of the mass. From a single particle or a collection of particles the direction of these photons in space would entirely random.

$$\lambda_c = \frac{\hbar}{m_1c + m_2c + m_3c +} \tag{1.9}$$

Gravitation and the Speed of Light

Vacuum Induced Index of Refraction

A number of researchers using the Schwinger electric density limit have investigated the index of refraction as it relates to pair production and, birefringence primarily at the lower limits of the energy density. It is certain that the velocity of light in the universe is related to vacuum polarization and thus it is important to understand the relation between mass and the velocity of light.

The vacuum polarization effect between two interaction photons is a well-researched process both from theoretical and experimental aspects. The first development by Sauter, Serber, Euler and others, [15][16][17], and later by more sophisticated methods of QFT by Schwinger and others[18][19].

The study of the vacuum polarization on the index of refraction is quite extensive in the lower levels of E when birefringence on photons in static fields effects are predominant, [20][21][22][23][24][25][26][27][28].

Urban [29] has proposed an index of refraction of space based on the vacuum polarization creating and annihilating electron pairs Within the uncertainty limit, In another publication Urban, [30], proposed an index of refraction depending on the time delay of a photon passing through the volume of intersecting particles Others have proposed field time delay operators that assert a delay photon delays [31][32][33][34]. This paper will propose that density of Feynman photons from the internal paths of massive particles sets the velocity of light and is the result of the probability of intersecting Feynman photons from the mass particles that inhabit the universe.

It is well known that a photon moving in a gravitational field has a trajectory that can be defined by Fermat's principle in Minkowski flat space with a variable speed of light with no other gravitational influence. The relation for the index of refraction developed from GR by Blandford, & Thorne, and others, with a flat metric is: [4], [35]

$$\eta^{-1} = \frac{c}{c_0} = \left(1 - \frac{2\mu}{r}\right) \;\rightarrow\; \frac{\Delta c}{c_0} = \frac{2\mu}{r} \qquad (1.10)$$

(this paper will refer to $\Delta c / c_0$ as an in Eq.(1.10), as an index of refraction.)

In previous papers [2], the author has illustrated that for photons, and confined light speed particles; a gradient in c produces the exact effect of gravitation on a massive particle with the same energy content. The illustration of a gradient in c generated by QFT equivalent to gravitation therefore creates a mechanism for gravitation within a Lorenz, local conservation of energy four-space.

The value Δc is the difference between the value of c a fixed point in space and the value induced by an intervening mechanism, generally induced by Feynman photons.

Logical Points

1: A photon trapped between two reflectors whether mirrors or in a black box cavity, represents rest mass. From the outside the photon is part of the total energy and cannot be ignored as part of the proper mass.

2: If a trapped photon is mass then a photon oscillating between two points must generate gravitation the same as a mass particle of the same energy.

Multiplying the change in c as a result of being in the probable location of a photon times the probability of a Feynman photon being at that point distant form its center, gives the change in c. (Postulate #3).

$$\Delta c = c_0 \frac{\lambdabar_{PL}^2}{\lambdabar_C^2} \frac{2\lambdabar_C}{r}$$

(1.11)

This simplifies to the change in c induced by gravitation

$$\frac{\Delta c}{2c} = \frac{\mu}{r}$$

(1.12)

Comparing the results of Eq.(1.12), with the well-established velocity of light induced by gravitation in flat Minkowski space Eq.,(1.10) The proposed probability of the Feynman photons, and the proposed change in the index of refraction induced by interaction with those photons establishes physical agreement with the gravitationally induced change in c.

The Ambient Speed of Light and Vacuum Density of the Universe

As can be seen from Postulate #1, the velocity of light is slowed by the probable presence of Feynman Photons. Every particle in the universe generates Feynman photons and the protons-neutrons mass are the largest constituents, and thus have by far the greatest effect of the ambient speed of light. Estimates of the mass and number of protons in the universe have been made, and can be used to approximation of the number density of Feynman photons. The speed of light in an empty universe would infinite, thus, the sum of the change and the current value is due to the probability density of Feynman photons, and the change predicted in postulate #1

From an estimate of the mass in the universe by D. Valev [39], which is the same as setting the radius of the visible universe as the Schwarzschild radius, the relation between the mass and the radius in a flat universe is:

$$R = \frac{2Gm}{c^2} \qquad \rightarrow \qquad \frac{Gm}{c^2(R/2)} = 1 \tag{1.13}$$

Most of the mass in the universe consists of proton mass particles then the sum of the number of those particles is:

$$n_P = \frac{R}{2\mu_P} = 5.2176112E+79 \quad \text{Proton equivalent masses} \tag{1.14}$$

This mass estimate in Eq. (1.13), can be written in the form of a sum of the particles in the postulates, Eq. (1.5), relating the velocity of light to the sum of the Feynman photons.

$$\frac{Gm}{c_0^2}\frac{\lambda}{\lambda^2}\frac{2\lambda}{R} = \sum_n \frac{\lambda_{PL}^2}{\lambda_P^2}\frac{(2\lambda_P)_n}{r_n} = 1 \;, \tag{1.15}$$

or:

$$c_0 \sum_n \frac{\lambda_{PL}^2}{\lambda_P^2}\frac{(2\lambda_P)_n}{r_n} = c_0 \tag{1.16}$$

This is a sum over all the particles in the universe and a single mass particle or group of particles can be separated such that:

$$c_0\left(\frac{\mu_1}{\lambda_1}\frac{(2\lambda_P)_1}{r_n}\right) + c_0 \sum_n \frac{\mu m c}{\hbar}\frac{(2\lambda_P)_n}{r_n} = c_0 \tag{1.17}$$

The second term is the speed of light in the universe; less the contribution of the first term, thus this term becomes c

$$c = c_0 \sum_n \frac{\lambda_{PL}^2}{\lambda_P^2}\frac{(2\lambda_P)_n}{r_n} \tag{1.18}$$

The value of c is the value of c_0 without the contribution from the particle or group of particles separated thus Eq.(1.17), becomes:

$$c_0 - c = = c_0 \left(\frac{\mu}{\lambda} \frac{(2\lambda_P)_n}{r_n} \right) \quad \rightarrow \quad \frac{\Delta c}{c_0} = \left(\frac{\mu}{\lambda} \frac{(2\lambda_P)_n}{r_n} \right) \quad (1.19)$$

Thus postulate #3, which is the change in the local index of refraction as a result of the Feynman photons, is the same as its change in the ambient index of refraction in the universe.

Feynman Photon Density in the Universe

The sum in Eq.(1.16), of the probability density ratio is just the density of Feynman photons in the universe, and can be found noting that the most of the contributions to the sum is due to the protons and thus can be factored out of Eq.(1.15). The ratio of the proton gravitational radius and the Compton radius is then gives the value of the sum, which is the number density of Feynman photons inhabiting the universe.

$$\frac{\lambda_P}{\mu_P} = \sum_n \frac{(2\lambda_P)_n}{r_n} = n_F \quad (1.20)$$

The sum is the total time average probability density of Feynman photons n_F at any given point in the universe, and the value is just the ratio of the Compton wavelength to the gravitational radius of the proton or, λ_P / μ_P

$$n_F = \frac{\lambda_P}{\mu_P} = 1.69321E + 38 \quad \text{Photons per cm}^3 \quad (1.21)$$

The ratio in Eq.(1.20), would be exact if the free particle protons-neutron mass constituted all the mass, but there are chemical, nuclear, and gravitational mass defects for particles in the universe that induce some difference.

Eq.(1.20), is the total probability number density of Feynman photons in the universe, but only half are responsible for any photon-photon interaction. The one-half reduction is the result of the Lorentz transform that prohibits the interaction of photons in the same direction [40], thus generally the interacting photons experience a flux density of just $n_F/2.$.

$$\frac{n_F}{2} = 8.4660E+37 \qquad (1.22)$$

The Feynman photon density Eq.(1.22), is thus the oncoming flux that a moving photon experiences inducing the index of refraction. (See endnote on the de Broglie-Bohm Pilot Wave Connection)

The Feynman photons thus set the index of refraction for the universe:

$$\frac{\Delta c}{c_0} = \frac{\bar{\mu}}{\bar{\lambda}} n_F = \frac{2\bar{\mu}}{\bar{\lambda}}\left(\frac{n_F}{2}\right) \qquad (1.23)$$

The bars indicate the average mass and wavelength of the individual mass particles in the universe. This density in Eq.(1.22), compares nearly exactly with the number density of photons in an electron in its classical volume which is:

$$\frac{2}{\left(\alpha\lambda_e\right)^3} = 8.9374E+37 \qquad (1.24)$$

Combining Eq.(1.21), and Eq.(1.24), the mass of the electron can then be approximated from the ratio of the gravitation radius and the Compton radius of the proton.

$$\lambda_e = \frac{1}{\alpha}\left(\frac{4\mu_P}{\lambda_P}\right)^{1/3} = 3.93216E-11 \text{ cm} \qquad (1.25)$$

The Feynman photon density n_F of the universe provides the index of refractions responsible for the effects attributed to the vacuum polarization defined by the Schwinger limit of electromagnetic energy density [41], and discussed in the section on electric scaling.

The Photon

The following is a proposed model that gives a physical interpretation of the result of the equations that induce the electrical particle-particle interaction. The graphical representation may not be the only possibility.

In earlier papers [13], there has been proposed a model of the electron consisting of Photons locked by vacuum polarization generated changes in the index of refraction. The purpose here is to present a plausible model of a photon that fits the known properties and allows calculations of physical values.

Presented here is the photon being a Planck size particle rotating at the Compton frequency inside the Compton volume. The Planck particle contains the spin of the photon. The Planck core has an inherent angular momentum of $\hbar$ that has a defined axis and is present even when the particle is not rotating. The rotation of the Planck particle defines the direction of a probability amplitude flow of where the Planck particle may be, but gives no contribution to the fixed $\hbar$ angular momentum.

The wave of the photon is regarded as the probability amplitude of its future location having no energy or energy density. When the particle transfers energy to another particle by Planck particle-particle interaction, the past probability amplitude of where it could be, vanishes everywhere instantly, and the future probability of where it may be is instantly created. The wavefunction, which is a probability of where it could be, collapses instantaneously on any interaction without a violation of special relativity.

Previously energy assigned to the photon has been thought of as an electromagnetic energy envelope, but as in the case of particle solution of the Schrodinger and Dirac Equation it has been demonstrated that the probability localization of the electron defined by the Photon Wave Equation is proportional to the electromagnetic energy density [37], [38]. The identification and replacement of the energy density of the electromagnetic field with the Planck particle probability flow amplitude is thus a scaling issue.

Model

As the Planck core rotates around its spin axis, there is radiated away in opposite directions probability flow amplitude of the future location of where the particle could be. This probability flow moves with c, and terminates at the Compton radius.

Figures 1, and 2 illustrates the revolving of the photon around the spin axis and the probability flow amplitude emanating from core.

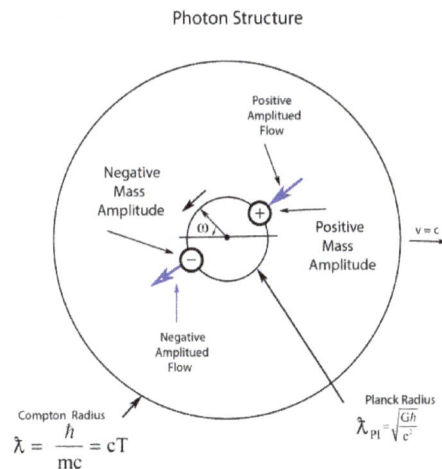

Photon Structure

Fig. 1

The flow amplitudes continuously radiate from the core at the velocity of light making the amplitude flow probability an Archimedes spiral. The maximum distance of this flow is the Compton radius since at that distance the source rotates back to its initial position and the flow at the Compton radius, which can't radiate must vanish.

The probability amplitude flows from the positive to the negative starting at the Compton radius and vanishing at the Compton radius as it flows out.

The spin vector of the Planck particle can have three orientations: forward, backward, and perpendicular to the path, thus the amplitude flow probability is polarized as; helictical, anti-helictical, and planar. If the rotation is transverse to the motion then the forward motion is a plane polarized wave, with a two state configuration of up and or down.

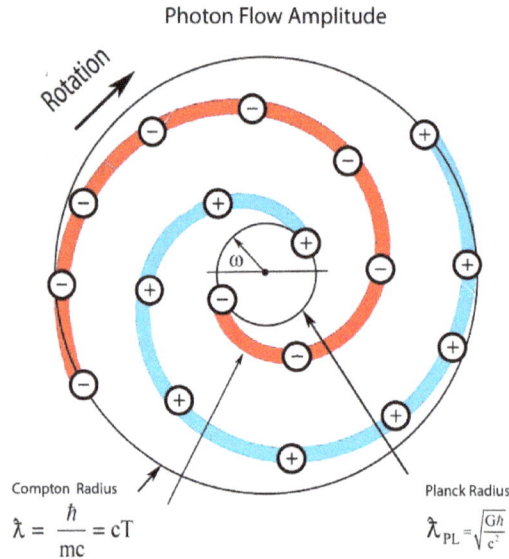

Fig. 2

Figure 2, shows the positive and negative flow amplitude originating at the rotating Planck particle, moving outward and terminating at the Compton radius when the angle returns to the original position. Probability flow from a Planck particle can't radiate beyond the Compton radius. The flow amplitude probability is the location the particle could intersect the probability flow amplitude of another photon and the product of the amplitudes determines the flow density and the direction of the oncoming photon.

Bound Radially polarized Photons

If the photons in Fig. 2 are moving around a circle, at the same frequency as the rotation of the photon, then the flow would be in a line constantly along the radial axis.

Inside the electron this product of the flow amplitude induces a flow density between the two orbiting photons and thus a radial index of refraction in the direction of the opposite photon.

On the exterior of the electron the Feynman photons probability flow amplitude are likewise polarized along the radius vector Fig. 3.

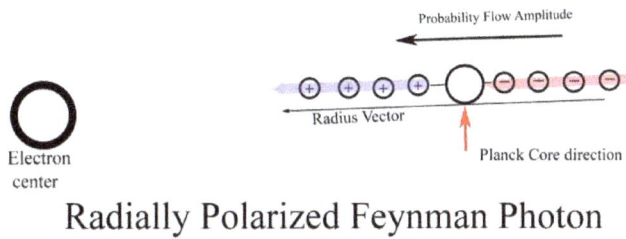

Radially Polarized Feynman Photon

Illustration of the flow probability amplitudes and direction of the Feynman photons resulting from the rotation of the photon around a center of momentum with the same frequency that it revolves around its core.

Fig. 3

The Feynman photons existing exterior to the core action path, maintain the radial polarization, on encountering photons from other particles.

On encountering a polarized photon from another electron or positron, the photon will see the photon flow density either positive or negative along the radial vector, causing the photon to be either attracted or repelled to the other particle.

The illustration of this interaction is shown in Fig. 4

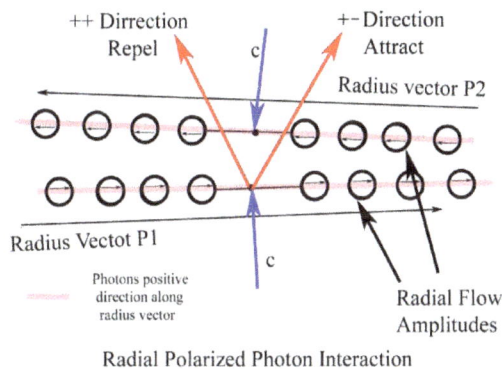

Radial Polarized Photon Interaction

Photon-Photon interaction

Fig.4.

When the Feynman photons from opposite particles meet at a point in space the probability flow amplitudes multiply to form a flow density. The relative signs determine the flow density for each oncoming photon.

In the electron the transverse probability amplitudes are opposite thus the flow is negative along the radius vector and the photons are bound together Fig. 5.

213

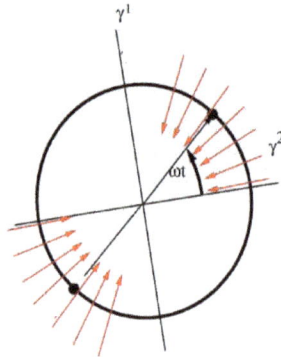

wwww

As probability amplitude of the exterior Feynman photons encounter similar photons from other charged particles there's a deflection either positively or negatively along its own radius vector. The product of the probability amplitudes which gives the flow density to the oncoming photon is the source of the charge effect of the particles.

As polarized photons from different electrons arrive at the same point in space, the product of the amplitudes determines the flow density direction. The effect is symmetric for both photons being attracted or repelled along their respective radius vectors

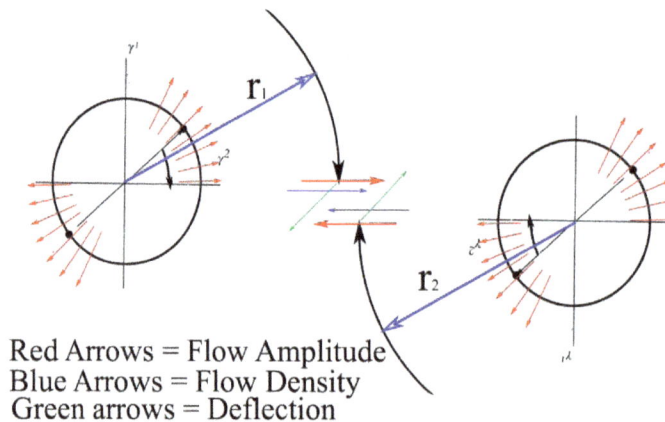

Red Arrows = Flow Amplitude
Blue Arrows = Flow Density
Green arrows = Deflection

Fig 6.

Charged Particle Interaction and the Value of the Fine Structure Constant

From the third postulate Eq.(1.5), the probability of a Feynman photon from an electron being at a point is:

$$\frac{\Delta c}{c} = \frac{\lambda_{PL}^2}{\lambda_e^2} \frac{2\lambda_e}{r} \tag{1.26}$$

The photons both in the electron orbit axially around the spin vector maintaining a ½ angular momentum, thus at any point in space the Feynman radially polarized photon repeats in the same direction and orientation as the action path of the orbit repeats. The repetition of this is then at the Compton frequency which increases the flow density at that point, Eq.(1.26), by v, the Compton frequency $v_e = \omega_e / 2\pi$.

Eq.(1.26), then becomes:

$$\frac{\Delta c}{c} = \frac{\lambda_{PL}^2}{\lambda_e^2} \frac{2\lambda_e}{r} v_e \tag{1.27}$$

or:

$$\Delta c = c_0 \left(\frac{\mu_e}{\lambda_e} \frac{2\lambda_e v_e}{r_1} \right) \tag{1.28}$$

This c_0 is the value of Δc at a distance r_1 from the origin of the particle.

If there is a second particle at a distance Δr_{21} from form the observation point then:

$$c = c_0 \left(1 - \frac{\mu_e}{\lambda_e} \frac{2\lambda_e v_e}{\Delta r_2} \right) \tag{1.29}$$

215

The c of the second particle at the observation point becomes c_0 at the observation point thus:

$$c = c_0\left(1 - \frac{\mu_e}{\lambda_e}\frac{2\lambda_e v_e}{\Delta r_2}\right)\left(\frac{\mu_e}{\lambda_e}\frac{2\lambda_e v_e}{r_1}\right) \qquad (1.30)$$

or:

$$\frac{\Delta c}{c_0} = \frac{\Delta c_1}{c_0} - \left(\frac{\mu_e}{\lambda_e}\frac{2\lambda_e v_e}{\Delta r_2}\right)\left(\frac{\mu_e}{\lambda_e}\frac{2\lambda_e v_e}{r_1}\right) \qquad (1.31)$$

The first term is the change in Δc due to the first particle. The second term is the interaction term between the particles dependent on the distance between the particles. This could be written:

$$\frac{\Delta c}{c_0} = \frac{\Delta c_1}{c_0} - \frac{\Delta c_I}{c_0} \qquad (1.32)$$

Where $\Delta c_I / c_0$ is now the change in the index of refraction due to the interaction of the particles thus:

$$\frac{\Delta c_I}{c_0} = \left(\frac{\mu_e}{\lambda_e}\frac{2\lambda_e v_e}{\Delta r_2}\right)\left(\frac{\mu_e}{\lambda_e}\frac{2\lambda_e v_e}{r_1}\right) \qquad (1.33)$$

Regrouping, and noting from the previous discussion on the interaction of radial polarized photons the product of this interaction has a $\pm$ sign:

$$\frac{\Delta c_I}{2c_0} = \pm\left(\frac{2\mu_e}{\lambda_e}v_e\right)^2 \frac{\lambda_e}{r_1}\frac{\lambda_e}{\Delta r_{12}} \qquad (1.34)$$

The quantity in brackets must be and is the Fine Structure Constant squared, (α^2), value of which is discussed and calculated in Appendix II. The merit of this theory hinges considerably on this fact, and since most of the terms in this expression are known to at least 10 significant digits there is very little room for error. The uncertainty in the value of the gravitational constant may be the most significant issue.

Eq. (1.34), can thus be written as:

$$\frac{\Delta c_I}{2c} = \frac{\alpha \lambdabar_{e1}}{r_1} \frac{\alpha \lambdabar_{e2}}{\Delta r_{12}} \qquad (1.35)$$

The value of the index of refraction in this expression is at the position from the first particle thus valuating the index of refraction at the classical radius $r_1 = \alpha \lambdabar_e$ of the first particle gives, the value function of the distance to the second particle. This is then the ratio of the potential energy to the total to the total energy of the particle, or the energy ratio ot the potential to the total energy of the electron.

$$\frac{\Delta c_I}{2c} = \frac{1}{m_e c^2} \frac{Q^2}{\Delta r_{12}} \qquad (1.36)$$

I appears that $r_1 = \alpha \lambdabar_e$ is the time average orbital radius of the two orbiting photons, and the time average density of the two particles is expressed in Eq. (1.24).

The interaction of two mass particles by gravitation

The gravitational interaction of two masses can be found using the same procedure for gravitation as charge, and Eq. (1.12).

The combined effect of the two particles at a point at a distance r from the fist point and that point being Δr_{12} from that point is:

$$\frac{\Delta c}{c_0} = \frac{\Delta c_1}{c_0} \frac{\Delta c_2}{c_0} \qquad (1.37)$$

The total index of refraction at r is then:

$$\frac{\Delta c_T}{c_0} = \frac{2\mu_1}{r_1} \frac{2\mu_2}{\Delta r_{21}} \qquad (1.38)$$

Since the value of the index of refraction is valid at any point it is convenient to pick a point a constant relative point independent of the relative motion

For the same reason as for the electric evaluation the evaluation point is set at the sum of the Schwarzschild radius.

$$|\vec{r}| = (2\mu_1 + 2\mu_2) \tag{1.39}$$

then

$$\frac{\Delta c}{c_0} = \frac{2\mu_1}{2(\mu_1 + \mu_2)} \frac{2\mu_2}{\Delta r_{12}} \tag{1.40}$$

Writing this out explicitly gives the index of refraction in terms of the particle separation thus:

$$\frac{\Delta c}{2c_0} = \frac{1}{(m_1 + m_2)c^2} \frac{Gm_1 m_2}{\Delta r_{12}} \tag{1.41}$$

This is the proper value of the ratio of the potential energy to total energy of two gravitating particles or the energy extractable by a change in relative position.

Energy and the Index of Refraction

The first point is to note that the gravitation and electric potentials for an electron in Eq.(1.12), and Eq.(1.27), are identical accept for the factor of the Compton frequency ν

$$\text{Gravity} \quad \frac{\Delta c_e}{c} = \frac{\mu_e}{\lambda_e} \frac{2\lambda_e}{r_1} \tag{1.42}$$

$$\text{Electric} \quad \frac{\Delta c_e}{c} = \frac{\mu_e}{\lambda_e} \frac{2\lambda_e}{r_1} \nu_e \tag{1.43}$$

The difference is in the fact that Δc for gravitation is the result of the probability amplitude of a direct hit of a Planck particle, whereas the electric change is the selective change in c along the radius

interaction is the probability amplitude of a Feynman photon having a direct hit with the tangential flow which is repetitive at a point in space at the Compton frequency.

For each of the potential relations, half the change in the index of refraction relative to the ambient value is equal to the ratio of the extractable or potential energy, to the total energy. The value of this can be understood in terms of the relation between potential and kinetic energy.

$$\frac{\Delta c}{2c_0} = \frac{1}{mc^2}\frac{Q^2}{\Delta r_{12}} \qquad \frac{\Delta c}{2c_0} = \frac{1}{(m_1+m_2)c^2}\frac{Gm_1m_2}{\Delta r_{12}} \qquad \frac{\Delta c}{2c_0} = \frac{\mu}{r} = \frac{Gmm}{mc^2r} \qquad (1.44)$$

The left side of these expressions is the ratio of the extractable potential energy for two particles to the total energy of the mass.

$$\frac{\Delta c}{2c_0} = \frac{\varepsilon_P}{\varepsilon} \qquad (1.45)$$

Identifying the index of refraction relation, Eq.(1.45), with the dynamics of mass and particle interaction is facilitated using the relativistic mass energy relation which can be expressed in the form:

$$\frac{(m-m_0)c^2}{mc^2} = \frac{1}{2}\frac{v^2}{c^2} \qquad (1.46)$$

The energy difference in Eq.(1.45), can be written as a change in the mass as a result of the change in c , That is:

$$\varepsilon_P = \frac{(m_0-m)c^2}{mc^2} \qquad (1.47)$$

From the Relativistic energy relation, the mass energy difference is the result of the kinetic energy, and in a conservative system these are just opposites, thus:

$$\frac{\Delta c}{2c_0} = \frac{\varepsilon_P}{\varepsilon} = \frac{(m_0-m)c^2}{m^2c^2} = -\frac{1}{2}\frac{v^2}{c^2} \qquad (1.48)$$

For a particle moving is space the change in the index of refraction of space is thus connected to the change in the kinetic energy.

Electric Field Scaling and the Schwinger Limit

Has been asserted here that the electric field is actually the future flow amplitude of photons, and as such it has no energy content. If this is true, then there must currently be an artificial energy scaling between the electric and probability densities. The assignment of a continuous energy field distribution to a probability distribution works well for engineering purposes and conforms to Maxwell's equations, but fails in theoretical applications. Historically this extractable energy been assumed to be the result of an energy in a field, that can be extracted by relative position of the charges

In order to have a consistent energy relation the energy density of the electric field is assigning to a point in space surrounding a charge has nominally been set to be:

$$E^2 = \frac{1}{8\pi}\frac{Q^2}{r^4},$$

(1.49)

This has been arrived at by integrating this density over all space, down to the classical radius of the electron and set it equal to the total energy.

$$\varepsilon_T = \int_{r_0}^{\infty} E^2\, 4\pi r^2 dr = \int_{r_0}^{\infty} \frac{1}{8\pi}\frac{Q^2}{r^4}\, 4\pi r^2 dr = \int_{r_0}^{\infty} \frac{Q^2}{r^2}\, dr = -\frac{Q^2}{r}\bigg|_{\alpha\lambda}^{\infty} = m_e c^2$$

(1.50)

Although this mechanically works it is arbitrary and obscures the actual mechanics of charge-charge interaction.

Charge Electric Field

We can look at this in terms of the ratio between the energy density and the Feynman photon density.

From the energy density perspective the potential from a charge is:

$$\frac{Q^2}{r} = \frac{\alpha c \hbar}{r} \tag{1.51}$$

At the Bohr distance from the charge the potential energy is:

$$\varepsilon_P = \frac{\alpha c \hbar}{\lambda / \alpha} = \alpha^2 mc^2 = 2\Re \tag{1.52}$$

Dividing this by the Rydberg energy gives:

$$P = \frac{2\lambda_e}{r} \rightarrow \frac{2\lambda}{\lambda} = 2 \tag{1.53}$$

From the density perspective this is the number density for the electron, or the number density of our postulate for the number of Feynman photons at the Compton radius. This leads to the presumption that the energy density defined for the electric field per Feynman photon is the Rydberg energy

Schwinger Energy Density Limit

Eq. (1.52), is the probability density at its lowest level of particle interaction. The Schwinger limit represents the other end or the highest level of particle interaction, and both can be shown to have the same scale factor between the electric and particle density.

From definitions the Schwinger electrical energy density limit and the Rydberg energy are:

$$E_S^2 = \frac{c\hbar}{\alpha \lambda_e^4} \qquad \Re = \frac{\alpha^2 mc^2}{2} \tag{1.54}$$

The ratio of the Schwinger energy density divided by the Rydberg gives exactly the classical volume of the electron,

$$\frac{E_S^2}{\Re} = \frac{2}{(\alpha \lambda_e)^3} = 8.9374E+37 \ \mathrm{cm}^{-3} \tag{1.55}$$

If as we have postulated the electron has two photons, then the Schwinger electric energy density is the Feynman photon density times one Rydberg of energy per photon

From Eq.(1.22), the number density of Feynman photons in the universe is

$$\frac{n_F}{2} = 8.466E+37 \text{ cm}^{-3} = \frac{2}{(\alpha \lambdabar_e)^3} \tag{1.56}$$

Multiplying it by the Rydberg constant gives the Schwinger limit.

$$E_{SW}^2 = \frac{n_F}{2} \Re = 1.94824E+27 \text{ erg cm}^{-3} \tag{1.57}$$

This is about 5% from the calculated value of the Schwinger Limit derived by Schwinger of 1.8454874E+27 ergs cm^{-3}

Both Eq.(1.53), and Eq.(1.57), lead to the presumption that the energy density ε_D, of the electric field has been defined arbitrarily equivalent to be the Feynman photon probability density times the Rydberg constant.

$$\varepsilon_D = n_F \Re \tag{1.58}$$

It is concluded that this represents the scaling relation between the Feynman photon probability amplitude and the electric field density.

*Only half of the Feynman photons in space can interact with another photon because Lorentz considerations prevent photons not having velocity components in the same direction from interacting

Conclusion

Presented has been a of a plausible causation of gravitation and electrical charge interaction within the confines Quantum Theory in Minkowski four-space, as well as a causation of the ambient velocity of light in the universe.

It has to be regarded as a new approach to physics, and as such a degree of speculation has been incorporated, many parts lack mathematical rigor, but fit well with known physical parameters. How the postulates and speculation impact or elucidate the understanding of QM will be left to future clarification.

Hopefully it will lead to a better understand between quantum mechanics and the internal dynamics of particles

References:

1. E. Noether's Discovery of the Deep Connection between Symmetries and Conservation Laws
Nina Byers, arXiv:physics/9807044v2 [physics.hist-ph] 23 Sep 1998

2 DT Froedge, "Gravitation is a Gradient in the Velocity of Light"
http://www.arxdtf.org/css/Gravitation.pdf, http://vixra.org/abs/1501.0052

3. Feynman, Hibbs, 1965, Quantum Mechanics and Path Integrals McGraw-Hill

4. Roger Blandford, Kip S. Thorne, Applications of Classical Physics, (in preparation, 2004), Chapter 26
http://pmaweb.caltech.edu/Courses/ph136/yr2012/1227.1.K.pdf

4 A. Kregar Aharonov-Bohm effect, http://mafija.fmf.uni-lj.si/seminar/files/2010_2011/seminar_aharonov.pdf

5. B. Smith, Photon wave functions, wave-packet quantization of light, and coherence theory, http://iopscience.iop.org/article/10.1088/1367-2630/9/11/414

6. S. Kocsis *et al.* Observing the average trajectories of single photons in a two-slit interferometer, Science. 2011 Jun 3;332(6034):1170-3. doi: 10.1126/science.1202218

7. Time-delay of classical and quantum scattering processes: a conceptual overview and a general definition Review Article
Massimiliano Sassoli de Bianchi1

8. Yi Liang and Andrzej Czarnecki Alberta Thy 12-11
Photon-photon scattering: a tutorial arXiv:1111.6126v2 [hep-ph] 8 Dec 2011

9. I, Bialynicki-Birulaphoton Photon Wave Function, Progress In Optics XXXVI, pp. 245-294 arXiv:quant-ph/0508202v1 26 Aug 2005

10. DT Froedge, A Physical Electron-Positron Model in Geometric Algebra, V041917 http://www.arxdtf.org/css/electron.pdf, http://vixra.org/pdf/1703.0274v2.pdf

11. M. Peskin, D. Schroeder, An Introduction To Quantum Field Theory. 1995 Westview press.

12. K, Bliokh, et.al, Photon trajectories, anomalous velocities, and weak measurements: A classical interpretation, New J. Phys. 15, 073022 (2013) , https://arxiv.org/abs/1304.1276

13. DT Froedge, Photon-Photon Vacuum Polarization Composite Electron Model

14. S Sakoda , M Omote, Difference in the Aharonov-Bohm Effect on Scattering States and Bound States Difference in the Aharonov-Bohm Effect on Scattering, Advances in Imaging and Electron Physics, v 110, 1999, Pages 101-171, Academic Press, 1999

15. Sauter, "Über das Verhalten eines Elektrons im homogenen elektrischen Feld nach der
relativistischen Theorie Diracs", Zeitschrift für Physik, 82 (1931) pp. 742–764.
doi:10.1007/BF01339461 (https://dx.doi.org/10.1007%2FBF01339461)

16. W. Heisenberg and H. Euler, "Folgerungen aus der Diracschen Theorie des Positrons",
Zeitschrift für Physik, 98 (1936) pp. 714-732. doi:10.1007/BF01343663 (https://dx.doi.org/10.1007%2FBF01343663) English translation (http://arxiv.org/abs/physics/0605038)

17 R. Serber, "Linear modifications in the Maxwell field equations," Phys. Rev. 48, 49 (1935); E. A. Uehling, Polarizationeffects in the positron theory, Phys. Rev. 48, 55-63 (1935).

18. J Schwinger, On Gauge Invariance and vacuum polarization , Physical Review volume e82, 1951

19. L. Rosenfeld Theory of Electrons (North Holland Publishing Company,
Amsterdam) 1951 arXiv:0906.3018v1 [physics.gen-ph]

20. W Dittrich, H. Gies, Vacuum Birefringence in Strong MagneticFields, arXiv:hep-ph/9806417v1 19 Jun 1998

21. T. Heinz M. Platz, Observation of Vacuum Birefringence: A Proposal
https://www.tpi.uni-ena.de/qfphysics/homepage/wipf/publications/papers/nonlineared.pdf

22. A. Rikken and C. Rizzo, Magnetoelectric birefringences of the quantum vacuum, Phys. Rev. A, Vol.63, 012107 (2000)

23. J. Heyl, L. Hernquist, Birefringence and Dichroism of the QED Vacuum
Lick Observatory, 12.20.Ds, 42.25.Lc 97.60.Jd, 98.70.Rz

24. T. Heinz, Observation of Vacuum Birefringence: A Proposal, https://www.tpi.uni-jena.de/qfphysics/homepage/wipf/publications/papers/nonlineared.pdf, 2005

25. W. Tsai, T. Erber, Propagation of photons in homogeneous magnetic fields. Index of refraction, Physical Review, D Volume 12, Number 4 1975.

26. C. Turtur, A Hypothesis for the Speed of Propagation of Light in electric and magnetic fields and the Planning of an Experiment for its Verification. Wolfenbüttel, November - 23 - 2007

27. E.Seegert, Quantum Reflection at Strong Magnetic Fields, Masterarbeit zur Erlangung des akademischen Grades Master of Science (M.Sc.).1986 in Berlin Matrikelnummer: Physikalisch-Astronomische Fakultät 2013

28. E. Zavattini, et. al., Experimental Observation of Optical Rotation Generated in Vacuum by a Magnetic Field, Phys. Rev. Lett 96, 110406 (2006)

29. M. Urban, F. Couchot, X. Sarazin, A. Djannati-Atai, The quantum vacuum as the origin of the speed of light, EPJ manuscript, arXiv:1302.6165v1 [physics.gen-ph] 2013

30. M. Urban A particle mechanism for the index of refraction, LAL 07-79, 2007,
Dmitri Kharzeeva and Kirill Tuchinb,Vacuum self–focussing of very intense laser beams, BNLNT-06/43, RBRC-657,arXiv:hep-ph/0611133v2 21 Feb 2007 https://arxiv.org/abs/0709.1550

31 Eisenbud-Wigner-Smith time (delay) operator Eisenbud (1948), Wigner (1955), Smith (1960)

32. E.P. Wigner, Phys. Rev. 98, 145 (1955). [39.26] F.T. Smith, Phys. Rev. 118, 349 (1960).

33. P. Brouwer, K. Frahm and C. Beenakker, Quantum mechanical time-delay matrix in chaotic scattering,
arXiv:chao-dyn/9705015v1 20 May 1997

34. M. deBianchi, Time-delay of classical and quantum scattering processes: a conceptual overview and a
general definition Review Article, Central European Journal of Physics, Volume 10, Number 2
(2012), 282-319 arXiv:1010.5329v3 [quant-ph] 2 Nov 2011

35. F. Karimi, S. Khorasani, Ray-tracing and Interferometry in Schwarzschild Geometry, arXiv:1001.2177
[gr-qc] arXiv:1206.1947v1 [gr-qc] 9 Jun 2012

36. S. Hossenfelder, Research Fellow Frankfurt Institute for Advanced Studies, discussion: The Planck length as a minimal length, http://backreaction.blogspot.com/2012/01/planck-length-as-minimal-length.html

37. B. Smith, Photon wave functions, wave-packet quantization of light, and coherence theory, http://iopscience.iop.org/article/10.1088/1367-2630/9/11

38. I, Bialynicki-Birulaphoton Photon Wave Function, Progress In Optics XXXVI, pp. 245-294 arXiv:quantph/0508202v1 26 Aug 2005

39. D. Valev, Estimations of total mass and energy of the universe, arXiv:1004.1035v1 [physics.gen-ph] 7 Apr 2010

40. P. Avery, Relativistic Kinematics II PHZ4390, Aug. 26, 2015, http://www.phys.ufl.edu/~avery/course/4390/f2015/lectures/relativistic_kinematics_2.pdf

41. J Schwinger, On Gauge Invariance and vacuum polarization , Physical Review volume e82, 1951

42 G. Gabrielse et al,, New Determination of the Fine Structure Constant from the Electron g Value and QED, Phys. Rev. Lett. 97, 030802 (2006) http://hussle.harvard.edu/~gabrielse/gabrielse/papers/2006/NewFineStructureConstant.pdf

43. T. Quinn et.al, The BIPM measurements of the Newtonian constant of gravitation, G https://royalsocietypublishing.org/doi/pdf/10.1098/rsta.2014.0032

44. Y. Aharonov, David Z. Albert, and Lev Vaidman How the result of a measurement of a component of the spin of a spin-1/2 particle can turn out to be 10, https://doi.org/10.1103/PhysRevLett.60.1351,

45. Y. Aharonov, L. Vaidman, Measurement of the Schrödinger wave of a single particle, Phys. Lett. A 178, 38 (1993).

46. R. Flack, et.al, Weak Measurement and its Experimental Realization, opscience.iop.org/article/10.1088/1742-6596/504/1/012016 arXiv:1408.5685v1 [quant-ph] 25 Aug 2014

47. Yi Liang and Andrzej Czarnecki Alberta Thy 12-11
Photon-photon scattering: a tutorial arXiv:1111.6126v2 [hep-ph] 8
Dec 2011

48 H. Gies, G. Torgrimsson Critical Schwinger pair production
Phys. Rev. D 95, 016001 – Published 23 January 2017 https://arxiv.
org/abs/1507.07802

49. R. Battesti, C.Rizzo., Magnetic and electric properties of
quantum vacuum. Reports on
Progress in Physics, IOP Publishing, 2013, in press

50. J. Heyl, L. Hernquist, Birefringence and Dichroism of the QED
Vacuum
Journal of Physics A: Mathematical and General, Volume 30,
Number 18 Nov. 1996

51. W. Tsai, Propagation of photons in homogeneous magnetic
fields: Index of refraction* Physical Review D Volume 12, Number
4 15 Aug. 1975

52. C. Doran, A. Lasenby 2003 Geometric Algebra for Physicists,
Cambridge University Press, Quantum Theory and Spinors, chapter
8

53. J. Evans, M. Rosenquist, *F=ma* Optics, American Journal of
Physics 54, 876 (1986); https://doi.org/10.1119/1.14861

54 C. Merkatas et. al.,Consensus Value for the Newtonian Constant
of Gravitation
Xiv:1905.09551v1 [physics.data-an] 23 May 2019

Appendix I

Probability Density of Off Path Feynman Photons

Though Feynman's proposal that a particle has a an equal probability of all paths, it is not true that the particle has an equal probability of being an any position at any distance from the path, in fact, there is no way at this time to directly calculate the probability amplitude as a function of the distance from the classical trajectory

For approximations one can turn to the work introduced by Aharonov, Albert, and Vaidman on "Weak measurements" [6], [12], [38], [45],]46], [47], that pre-selects an initial state, a measuring device, and a post-selected final state. The results that can be measured as well as calculated can yield approximations regarding the probability density as a function of distance from the classical trajectory

K. Bliokh *et al.* [12], extending the work of Kocsis *et al,* [6], using the quantum weak-measurements method introduced by Aharonov *et al.* [45], made measurements of the "average trajectories of single photons" in a two-slit interference experiment.

The "Weak Values", method implies averaging over many events, i.e., the same as a multi-photon limit of classical linear optics, and applicable to the multiple path of a reciprocating photon. Bliokh was able to give a classical-optics interpretation to the experiment, and asserted that weak measurements of the local momentum of photons made by Kocsis *et al.* [6], represent measurements represented an average over many events and thus the measurements of the Poynting vector in an optical field.

Bliokh found that the transverse location probability density for a Feynman photon as a function of radius form a Feynman path to be proportional to 1/r thus [12]:

Bliokh

The value of k in Eq.(3.2), is not found by the properties of the path integrals near the classical tack, and are not well understood even with the weak theory & weak measurements, but Sakoda and Omote [11] did calculated the differential cross section from the scattering amplitudes finding the asymptotic probability distribution $r \gg \lambda$ to be proportional to:

$$\mathbf{P}(r_\perp) = \psi^* \psi \to \lambda / r \qquad (1.59)$$

For the consideration of a trapped photon oscillating between two points or in circular motion Fig. 1a, bound by vacuum polarization induced gradient in the index of refraction, the path goes from one point to another and then back, thus doubling the probability density. The relative time average probability density of the Feynman photons being at a position at a distance r from the classical tract is postulated to be:

$$P_F = \frac{2\lambda}{r} \qquad (1.60)$$

Appendix II

Calculated Value of the Fine Structure Constant

DT Froedge 06-02-19

In Section X of the paper, Eq.(1.34), there is a value of α^2 found in fundamental constants that can be useful in testing the theory.

$$\alpha^2 = \frac{2\lambda_{PL}^2}{\lambda_e^2} v^2 \tag{2.1}$$

This is the value of the fine structure constant, Alpha (α), in terms of fundamental constants that are known to a high degree of precision by calculations from QFT considerations. The λ_e is the Compton radius of the electron, and v is the Compton frequency $v = c/(2\pi\lambda_c)$. The Planck length is:

$$\lambda_{PL}^2 = \frac{G\hbar}{c^3} \tag{2.2}$$

The value of α^2 is nominally:

$$\alpha^2 = \left(1/137\right)^2 \tag{2.3}$$

or more accurately from the work of Gabrielse et.al.[42],

$$\alpha = 0.00729735253594 \tag{2.4}$$

Whether this theory has merit or not, hinges to a degree on the accuracy of the prediction of this relation. This value of α (Eq.(2.1)), is expressed in fundamental constants known to a high accuracy (~12 significant digits) and there are no "fudge" factors, The major uncertainty in the calculated value is the value of the gravitational constant G, which is experimentally determined to only about 5 significant digits, is the major uncertainty. the gravitational constant..

A more explicit for Eq.(2.1), is:

$$\alpha^2 = \frac{G\hbar}{c^3} \frac{2}{\lambda_e^2} \frac{c^2}{2^2\pi^2\lambda_e^2} == \frac{G\hbar}{2\pi^2 c\lambda_e^4} \tag{2.5}$$

The model for the electron that generates the relation for α is of two photons revolving round the center of momentum, and just as an in the case of an electron revolving around the proton, the quantum loops of the Feynman paths of the orbiting photons must be taken into account. The mass calculated Compton wavelength is the first loop, and the effect on the wavelength of all the other loops must be taken into account, this is done by multiplying the Compton radius λ_e, by the anomalous gyromagnetic ratio $g_e / 2$

$$g_e / 2 = 1.0011596522 \qquad (2.6)$$

To assess the predicted theoretical value, the expression Eq. (2.5), can be solved for the gravitational constant, and compared with current experimental values.

Solving for the electron mass in terms of fundamental constants gives:

$$m_e = \frac{\hbar g_e}{c} \left(\frac{2\pi^2 c \alpha^2}{G \hbar} \right)^{1/4} \qquad (2.7)$$

Solving, for the gravitational constant gives:

$$G = \frac{\alpha^2 2\pi^2 c \left(\lambda_e g_e \right)^4}{\hbar} \qquad (2.8)$$

Calculating G from Eq. (2.8), to 12 significant digits gives:

$G = \underline{6.67586}727955 \times 10\text{-}08 \, \text{cm}^3 \text{gm}^{-1}\text{s}^{-2}$

This is slightly higher (0.02%) than the current Codata consensus recommended value of $6.67430(15) \times 10$ -11 $\text{m}^3\text{kg}^{-1}\text{s}^{-2}$, but it is not outside the scatter of current measurements, and is exactly on the Cavendish balance measured International Bureau of Weights and Measures, BIPM value by T. Quinn et.al, [43], published in 2015.

$G = 6.67586(36) \times 10\text{-}11$

The calculated value is in agreement with this value to the limits of its experimental accuracy of this measurement.

Accept for some numerology associations, the above relation Eq., is the only known physical relation between the fine structure constant, the Gravitational constant, and the mass of the electron.

Reference values:

Constants used in calculation in CGS units.

$c =$ 2.9979245800E+10 $g_F = g_e / 2 =$ 1.00115965218

$\hbar =$ 1.0545918473E-27 $\lambda_e = \hbar / (m_e c) =$ 3.8616633678E-11

$\alpha =$ 1/137.03599971 $m_e =$ 9.109389966E-28

Summary:

Calculated value of G $G = 6.67586727955 \times 10\text{-}11 \ \mathrm{m^3 kg^{-1} s^{-2}}$

BIPM Cavendish balance value of $G = 6.67586(36) \times 10\text{-}11 \ \mathrm{m^3 kg^{-1} s^{-2}}$

Codata Consensus Value $G = 6.67430(15) \times 10\text{-}11 \ \mathrm{m^3 kg^{-1} s^{-2}}$

Excerpt from BPIM report: *https://royalsocietypublishing.org/doi/pdf/10.1098/rsta.2014.0032*

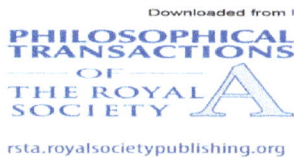

PHILOSOPHICAL TRANSACTIONS OF THE ROYAL SOCIETY

rsta.royalsocietypublishing.org

The BIPM measurements of the Newtonian constant of gravitation, G

Terry Quinn[1,†], Clive Speake[2], Harold Parks[1,‡] and Richard Davis[1,§]

12. A value for Newton's constant of gravitation

The peak-to-peak servo torque, τ_s, obtained as an unweighted mean of 10 data runs was $3.148869(94) \times 10^{-8}$ N m and using equations (9.2a) and (11.3) we can write

$$G_s = \frac{\tau_s}{\Gamma_s} = 6.67515(41) \times 10^{-11} \ \mathrm{m^3 \, kg^{-1} \, s^{-2}} \ (61 \, \mathrm{ppm}). \tag{12.1a}$$

The unweighted mean of the 10 data runs giving a value of the peak-to-peak deflection angle of 0.1529322(29) mrad using equations (9.2b), (8.2) and (11.8) we can write

$$G_c = \frac{\tau_c}{\Gamma_c} = 6.67586(36) \times 10^{-11} \ \mathrm{m^3 \, kg^{-1} \, s^{-2}} \ (54 \, \mathrm{ppm}). \tag{12.1b}$$

We have used the values for the uncertainties in the experimental measurements given in §11.

Experimental Data

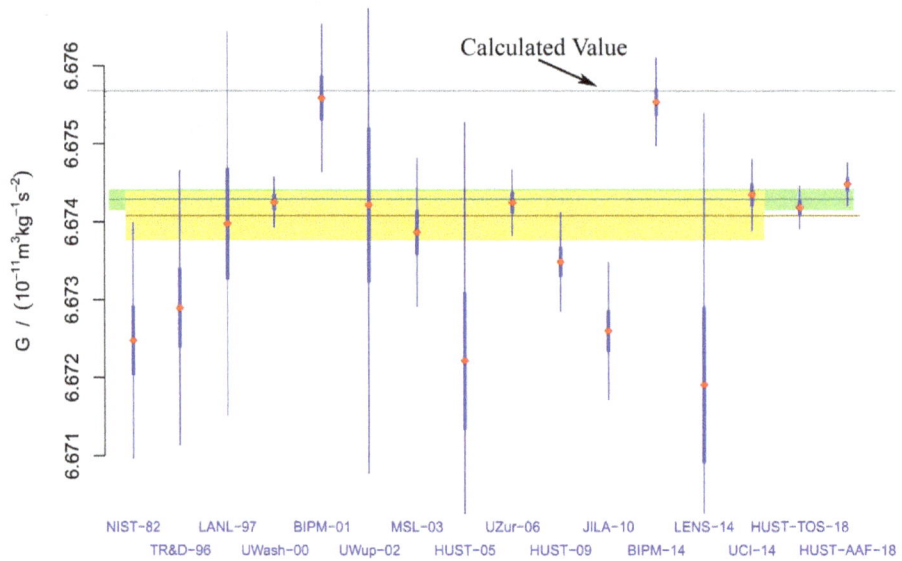

Experimental values used for statistical calculation of Codata consensus value of G [53]

Appendix III

Summary of Electron Model

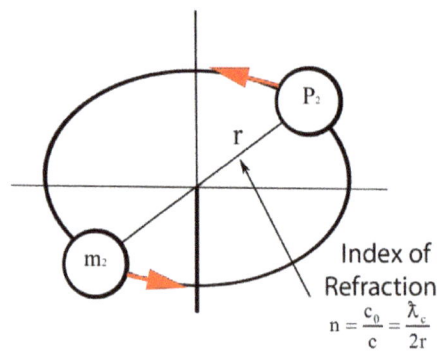

Orbiting Photons

Fig. 1a.

In an earlier paper "The Electron as a Composition of Two Vacuum Polarization Confined Photons" [13] the author developed the model of an electron as two bound photons in radially polarized along the orbital radius. Work done by a number of research-ers on the index of refraction at electromagnetic energies near the Schwinger limit, [41], have shown that the index of refraction as the result of the nonlinearity of the vacuum polarization is substantial . There is no doubt that the index of refraction at the Schwinger limit, [48],[49],[50],[51], is be sufficient to induce stable circular mo-tion, and there is no doubt that single photons can have that density.

The general form of the index of refraction developed from that work is:

$$\frac{c}{c_0} = \left(1 - k\frac{\alpha^2 E^2}{E_{SW}^2}\right) \quad \rightarrow \quad \frac{\Delta c}{c_0} = k\frac{\alpha^2 E^2}{E_{SW}^2} \tag{3.1}$$

By the Lorentz transformation, manifest in the Thomas preces-sion, an inertial frame rotating around a central point rotates about its path. If there is a helictical polarized photon moving in a circu-lar orbit in a circular index of refraction, the helictical rotation rate decreases proportional to the radius of the rotation. As the radius decreases to the Compton radius, the helictical rotation is stopped leaving the polarization continuously along the radial axis at one cycle per two revolution, in agreement with spinor considerations [52].

That is from an electrical perspective the electric field is radial about a point, and from the electrical perspective this would be a charged particle. From the perspective presented here this would be a probability of flow along the radial axis.

From classical mechanics it is known that a stable orbit for a pho-ton requires an index of refraction proportional to 1/r, [53]. Thus:

$$\frac{c_0}{c} = \frac{k}{r} \quad \rightarrow \quad r = \frac{kc}{c_0} \tag{3.2}$$

The angular momentum, the sum of the angular momentum of the two photons in the electron must be 1/2 thus:

$$L = rm_e c = \frac{kc}{c_0}mc = \frac{\hbar}{2} \tag{3.3}$$

From the postulates energy is an invariant of c thus:

$$m_e c^2 = \hbar \omega_e \tag{3.4}$$

Then Eq. (3.3), five the value of k to be:

$$\frac{k}{c_0} \hbar \omega_e = \frac{\hbar}{2} \tag{3.5}$$

and:

$$k = \frac{c_0}{2\omega_e} = \frac{\lambda_e}{2} \tag{3.6}$$

Putting this into Eq. (3.2), gives:

$$\frac{c}{2r} = \frac{c_0}{\lambda_e} = \omega_e \tag{3.7}$$

Since ω_e is a constant the electron orbits are stable and the angular momentum is, $1/2\hbar$.

A Quantum Field Theory Conjecture for the Origin of Gravitation

D.T. Froedge

V100418

Abstract

Gravitation defined in curved space has never been found to be compatible with quantum mechanics or quantum field theory. This is likely due to the fact that one theory is based in local conservation of energy and the other defines energy globally, and not locally conserved [Noether 1]. Equations mixed with variables from the two conservation laws, could neither be invariant nor covariant under coordinate transformations. This paper presents a theory for a gradient in c induced by QFT equivalent to the gradient in c induced by gravitation. In previous papers the author has illustrated that for photons, and confined light speed particles, a gradient in c demonstrates the effect of gravitation. The illustration of a gradient in c generated by Quantum Field Theory equivalent to gravitation therefore creates a theoretical mechanism for gravitation within a Lorenz, local conservation four-space.

With a few assumptions regarding the nature of photons and the reality of path integrals, a gradient in c equivalent to gravitation can be illustrated as a feature of Quantum Field Theory.

Introduction

Paul Davies in his introduction to *Six Easy Pieces* by Richard P. Feynman said:

"You could not imagine the sum-over-histories picture being true for a part of nature and untrue for another part. You could not imagine it being true for electrons and untrue for gravity"

If gravitation is a gradient in c as discussed by the author in other papers, then there must be a mechanism for inducing a change induced by a locally confined energy. This paper discusses how this is possible.

Feynman proposed that for a photon, or any particle, going from one point to another, there is a probability of the particle has traveled every possible path [2], and by very accurate measurements of quantum effects there is every reason to believe that this is true. It is not unreasonable to presume that the interaction of these photons with passing photons could make velocity changes to the index of refraction for these photons

Some say that the sum-over-histories picture is just a mathematical equivalence of QFT that predicts the proper path, and not really a probability of the particle being elsewhere. The action at distance of phenomena of the bell inequality, and the Aharonov-Bohm Effect [3], suggest a reality to the many path view, and for our purpose it will be proposed that the photons are real, near point particles that have a probability density located off the classical trajectory.

Speed of Light In Gravity

It is well known that a photon moving in a gravitational field has a trajectory that can be defined by Fermat's principle in Minkowski flat space with a variable speed of light with no other gravitational influence. The relation for the index of refraction developed from GR by Blandford, & Thorne with a flat metric is: [4], [5]

$$\eta^{-1} = \left(1 - \frac{2\mu}{r}\right) \quad \text{or} \quad c = c_0\left(1 - \frac{2\mu}{r}\right) \quad \text{or} \quad \frac{\Delta c}{c_0} = \frac{2\mu}{r} \tag{1}$$

The confined lightspeed sub-particles in a massive particle functions inertially equivalent to a mass particle and experiences the same acceleration in a variable index of refraction as a mass particle in a gravitational field. It has been argued that the internal constituents of all mass propagate with a velocity dependent on c and are also accelerated in a gradient just as confined photons. The same acceleration of massive particles as for photons would be induced by a gradient in c. [6]

Topics:

Included Appendix

Simplest Rest Mass Model

The photon carries an energy that, though in general tiny, must exert a gravitational pull on the particle whose position we wish to measure, and therefore must generate gravitation [7]. In order to define a simple thought experiment, the start will be by setting forth the simplest form of rest mass possible: a standing wave photon oscillation between two reflectors. This photon is functionally equivalent to a rest mass and must generate gravitation proportional to its confined energy $m = E/c^2$. The mechanism that induces gravitation must be present in this simple system, but there is very little in classical physics that would suggest a causal connection between the oscillating photon and the passing photon. The interaction induced by the conjecture of Feynman, of the photon paths taken by a particle going from one point to another existing outside the classical path could offer the causal connection.

Feynman Photons

From the Path integral formulation of QFT by Feynman, for a photons moving from point a, to point b there is a probability of existence outside the classical path [2], and as such there should be a probability of interaction with passing photons. With a few assumptions it will be shown that the change in c induced in the passing photons could be equivalent to the change in c induced by gravitation. Note that the photons discussed here are "not" an "off shelf" or virtual photon, but the real probable presence of the fully energetic photon existing throughout space. For the purpose of this paper, these photons will be referred to as "Feynman photons"

As an illustration Fig.1, shows a single photon oscillating as a standing wave between two points with the approach of an interloping external photon. On each cycle there is a new set of paths going both ways, and thus there is a multi trip averaging of trajectories for a single photon. Lorentz principles allow only free photons having velocity components in opposite directions to interact [8].

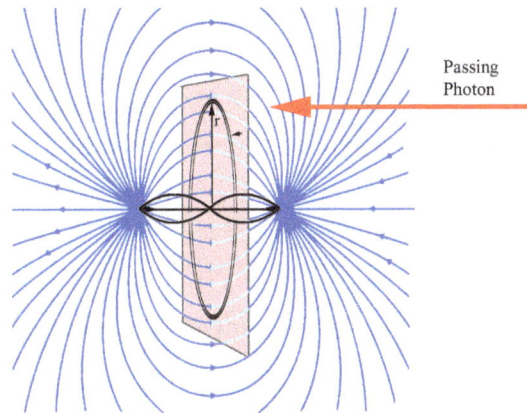

Passing
Photon

some of the possible
Feynman paths for a
photon oscilating between
two points.

Fig 1

Though Feynman's proposal that a particle has a an equal probability of all paths, it is not true that the particle has an equal probability of being at any position at any distance from the path, in fact, there is no way at this time to directly calculate the probability amplitude as a function of the distance from the classical trajectory [9]. This is a central problem to the issue and accurate results await further developments in QFT

For approximate solutions one can turn to the work introduced by Aharonov, Albert, and Vaidman on "Weak measurements" [10-11] that pre-selects an initial state, a measuring device, and a post-selected final state. The results that can be measured as well as calculated can yield approximations regarding the probability density as a function of distance from the classical trajectory. A presentation of this was by K, Bliokh, et.al [12], showing a radial probability density proportional to 1/r

$$\psi * \psi = \mathbf{P}\left(\mathbf{r}_\perp\right) = \frac{k}{r} \qquad (2)$$

Nature of the Photon

The photon, the first of the discovered particles is still not well understood. It contains a quanta of energy and although the wavelength can be thousands of miles long, it can deliver that energy to a single atom or electron instantly.

Because of the energy assigned to the electromagnetic wave the photon is mostly thought of as being the electromagnetic envelope, but as in the case of particle solution of the Schrodinger and Dirac Equation it has been demonstrated that the probability localization of the electron defined by the Photon Wave Equation is interchangeable with the electromagnetic energy density [13], [14]. Thus the identification and replacement of the energy density of the electromagnetic field with the photon location probability density is well justified. The energy of the photon is demonstrably not distributed over the entire wave but is localized well enough to be transferred instantaneously to a point particle. The distribution of the energy over distances of kilometers would make the instantaneous transfer of the energy a violation of special relativity.

If the electromagnetic envelope is just the probability envelope of location it is apparent that the instantaneous transfer of energy requires the physical energy carrier of the photon to be very small.

Propositions

1

The change in the speed of light on passing through a volume of space containing photons is proportional to the probability of a collision with a photon in that volume.

From Eq.(1), this is:

$$\frac{\Delta c}{c_0} = \Delta P \rightarrow \int n\sigma \; dx \qquad (3)$$

In the integral, n is the number density and is σ the photon-photon cross section.

2

The Energy and momentum of a photon is contained within the Planck radius. $\lambdabar_{PL}$. The transfer of energy and momentum by a photon is by the probability of interaction with its Planck cross section $\lambdabar^2_{PL} = \sigma_{PL}$

3

The electromagnetic wave associated with the photon is only the probability amplitude of the location of the photon having a radius of the Compton radius $\lambdabar_{PH}$, and having no energy density.

Specifics

Photon location probability radius:

$$\lambdabar_{PH} \tag{4}$$

Photon radius & Plank length definition:

$$r_{Ph} = \lambdabar_{PL} = \sqrt{\frac{G\hbar}{c^3}} \qquad\qquad \lambdabar^2_{PL} = \mu_{Ph}\lambdabar_{Ph} \tag{5}$$

The cross section of the photon:

$$\sigma_{Ph} = \lambdabar^2_{PL} = \frac{G\hbar}{c^3} \tag{6}$$

243

The energy density for the photon within the Planck volume:

$$\rho_{Ph} = \frac{\hbar\omega}{\lambdabar_{PL}^3} = \frac{\hbar c}{\lambdabar_{PL}^3 \lambdabar_{PH}} \tag{7}$$

4 Corollary:

If a photon is passing within the Compton radius of a second photon, the probability for interaction with the passing particle is proportional to the ratio of the cross section of the Planck particle to the Compton area.

$$P_{\gamma\gamma} = \frac{\lambdabar_{PL}^2}{\lambdabar_{Ph}^2} = \left(\frac{\sigma_{PL}}{\sigma_{Ph}} \right) \tag{8}$$

Rational for Postulates

Proposition 1

Urban's paper "particle mechanism for the index of refraction", [15] has postulated a similar relation in relation to a corpuscular mechanism for the index of refraction

$$[n-1] = N_{mol}\alpha\sigma_\perp c\Delta t \tag{9}$$

The number density is N_{mol}, $\alpha\sigma_\perp$ is the cross-section of the particle and Δt is the time delay in crossing the particle.

This refraction change is just the ratio of hitting the particle with a cross-section $\alpha\sigma_\perp$ within a distance of $c\Delta t$. (Note: $\Delta c / c_0 = [n-1]$)

This is an extension of the proposition of Urban that:

(The photon) has no spatial extension or, at least, it is smaller than the atomic dimensions. There is no frequency, no wavelength and no electric field associated with our photon.

We have modified this to assert that the small size of the photon is the Planck size particle with all photons having the same size, and the energy being contained within the Planck volume not in the electromagnetic field.

Photon-Photon Interaction

We can modify Urban's relation for a Planck size photon-photon interaction. The particle being impacted is another photon so first we can focus on the change in velocity on entering its probability distribution. On entering the Compton radius of a second Planck size particle the probability of interaction as expressed in Eq.(8), is the ratio of the cross section times the wavelength thus:

$$[n-1] = \frac{\Delta c}{c_0} = \frac{\sigma}{V}\lambda = \frac{P_{PL}^2}{\lambda_{Ph}^2}\lambda_{Ph} = \frac{\mu_{Ph}}{\lambda_{Ph}} \qquad (10)$$

This turns out to be the ratio of the gravitational radius to the wavelength, and is the change in velocity or refraction of the first particle inside the Compton radius of the second particle.

Multiplying Eq.(10), by the density of photons, or the density of other photons in a volume of space, gives the space refraction.

Probability of Feynman path photon Density

If the probability density of the Feynman photons surrounding the path of the oscillating photons noted above can be found then with Eq.(10), the equivalent gradient in c that is induced by gravitation can be found.

There is currently no direct way of calculation the probability density of the Feynman particle path or particle density. By the use of methods developed by Aharonov, Albert, and Vaidman related to "Weak Values" and "weak measurements", there is however some theoretical as well as experimental indication of the photon probability density associated with the actions paths.

Although there are many that have worked on this particular issue a definitive theoretical result awaits further development. This paper will take note of the theoretical and experimental work of K. Bliokh et al. [12] in arrival at a radial probability density. There are others that have researched this issue, but this paper is on point.

K. Bliokh et al. extending the work of Kocsis et al, [12], using the quantum weak-measurements method introduced by Aharonov et al. [11], made measurements of the "average trajectories of single photons" in a two-slit interference experiment.

The "Weak Values", method implies averaging over many events, i.e., the same as a multi-photon limit of classical linear optics, and applicable to the multiple path of a reciprocating photon. Bliokh was able to give a classical-optics interpretation to the experiment, and asserted that weak measurements of the local momentum of photons made by Kocsis et al. [24], represent measurements represented an average over many events and thus the measurements of the Poynting vector in an optical field.

Bliokh found that the transverse location probability density for a Feynman photon as a function of radius form a Feynman path to be proportional to 1/r thus:

$$\mathbf{P}(r_\perp) = \psi^*\psi \rightarrow k/r \tag{11}$$

The value of k in Eq.(11), is not found by the properties of the path integrals near the classical tack, and are not well understood even with the weak theory & weak measurements, and the relation could only be valid for $r \gg \lambda_{Ph}$ values of. The actual value must near the classical tract, is not linear in r, it has a higher probability in the radius of λ_{Ph}, and the integral over all space cannot exceed one.

For our purposes we will consider the value of k at large distances from the oscillating path. For the change in c induced by the Path Integrals from the trapped photon to match the change in c induced by gravitation, the value of k must be set to $k = 2\lambda_{PH}$, thus the probability density of a Feynman path photons being at a distance from the classical track of an oscillating photon as the interloping photon passes would have to be:

$$P_F = \frac{2\lambda}{r} \qquad (12)$$

The value of the constant: 2λ is the result of reverse engineering the need to match the effect of gravitation, and not a derivation of QFT. This is not an unreasonable or unexpected value, but more developed methods in QFT will be necessary to accurately evaluate this constant.

Spatial Index of Refraction induced by QFT

Multiplying the refraction change as the result of encountering a photon (Eq. (10)), by the probability of encountering a Feynman photon (Eq. (12)), gives the space refraction surrounding a quantity of mass to be:

$$\frac{\Delta c}{c_0} = \frac{\lambda_{PL}^2}{\lambda_{Ph}^2} \frac{2\lambda_{Ph}}{r} = \frac{2\mu}{r} \qquad (13)$$

This is the same change in c as induced by gravitation in Eq. (1).

This is the same as relationship between space, and mass as defined in General Relativity by the Einstein tensor, with the Einstein tensor providing the curvature of space as the source of gravitation. The relations developed in this paper provide the mechanism for gravitation as a change in the index of refraction resulting from QFT in four-space.

247

Conclusion

General Relativity is a curved multi-space construct without energy localization, and has not yet been proven to be compatible with QFT. The preceding development has shown a path to connect gravitation, and QFT through principles totally consistent with Quantum Mechanics, and with methodology in the confines of QFT and Four-space.

The assumptions regarding the nature the photon and its interaction with other photons are not out of bounds with experimental evidence, and the interaction of photons with probability amplitudes fits what is known of quantum interactions.

The postulation of the energy carrying kernel of the photon being the size of the Planck particle is a little unusual, but it fits better than the electromagnetic wave being the energy carrier. This is true both from the perspective of special relativity and quantum mechanics.

The absence of accurate QFT calculations of the Feynman particle probability amplitudes creates an obstacle to ultimate accuracy, but few would doubt that there is Feynman particle density, and that it would have an effect on passing photons.

This proposed theory allows the photon not only be the gauge boson and force carrier for electromagnetism but also to be the force carrier of gravitation.

There is some reverse engineering in the process that provides values consistent with known physics, (Eq.(12)), but the values necessary are not unreasonable.

References:

1. E. Noether's Discovery of the Deep Connection between Symmetries and Conservation Laws
Nina Byers, arXiv:physics/9807044v2 [physics.hist-ph] 23 Sep 1998

2. Feynman, Hibbs, 1965, Quantum Mechanics and Path Integrals
McGraw-Hill

3. A. Kregar Aharonov-Bohm effect, http://mafija.fmf.uni-lj.si/seminar/files/2010_2011/seminar_aharonov.pdf

4. Roger Blandford, Kip S. Thorne, Applications of Classical Physics, (in preparation, 2004), Chapter 26 http://pmaweb.caltech.edu/Courses/ph136/yr2012/1227.1.K.pdf

5. F. Karimi, S. Khorasani, Ray-tracing and Interferometry in Schwarzschild Geometry, arXiv:1001.2177 [gr-qc] arXiv:1206.1947v1 [gr-qc] 9 Jun 2012

6. DT Froedge, "Gravitation is a Gradient in the Velocity of Light" http://www.arxdtf.org/css/Gravitation.pdf, http://vixra.org/abs/1501.0052

7. S. Hossenfelder, Research Fellow Frankfurt Institute for Advanced Studies, discussion: The Planck length as a minimal length, http://backreaction.blogspot.com/2012/01/planck-length-as-minimal-length.html

8. P. Avery, Relativistic Kinematics II PHZ4390, Aug. 26, 2015, http://www.phys.ufl.edu/~avery/course/4390/f2015/lectures/relativistic_kinematics_2.pdf

9. Personal communication.

10. Y. Aharonov, D. Albert, and L. Vaidman How the result of a measurement of a component of the spin of a spin-1/2 particle can turn out to be 10, https://doi.org/10.1103/PhysRevLett.60.1351,

11. Y. Aharonov, L. Vaidman, Measurement of the Schrödinger wave of a single particle, Phys. Lett. A 178, 38 (1993).

12. K, Bliokh, et.al, Photon trajectories, anomalous velocities, and weak measurements: A classical interpretation, New J. Phys. 15, 073022 (2013) , https://arxiv.org/abs/1304.1276

13. I, Bialynicki-Birulaphoton Photon Wave Function, Progress In Optics XXXVI, pp. 245-294 arXiv:quant-ph/0508202v1 26 Aug 2005

14. B. Smith, Photon wave functions, wave-packet quantization of light, and coherence theory, http://iopscience.iop.org/article/10.1088/1367-2630/9/11/414

15. M. Urban, A particle mechanism for the index of refraction, 2007, https://arxiv.org/abs/0709.1550

16. Eisenbud-Wigner-Smith time (delay) operator Eisenbud (1948), Wigner (1955), Smith (1960)

17. E.P. Wigner, Phys. Rev. 98, 145 (1955). [39.26] F.T. Smith, Phys. Rev. 118, 349 (1960).

18. E.P. Wigner, Phys. Rev. 98, 145 (1955). [39.26] F.T. Smith, Phys. Rev. 118, 349 (1960).

19. Time-delay of classical and quantum scattering processes: a conceptual overview and a general definition Review Article Massimiliano Sassoli de Bianchi1

20. P. Brouwer, K. Frahm and C. Beenakker, Quantum mechanical time-delay matrix in chaotic scattering, arXiv:chao-dyn/9705015v1 20 May 1997

21. M. deBianchi, Time-delay of classical and quantum scattering processes: a conceptual overview and a general definition Review Article, Central European Journal of Physics, Volume 10, Number 2 (2012), 282-319
arXiv:1010.5329v3 [quant-ph] 2 Nov 2011

22. Yi Liang and Andrzej Czarnecki Alberta Thy 12-11 Photon-photon scattering: a tutorial arXiv:1111.6126v2 [hep-ph] 8 Dec 2011

23. R. Battesti, C. Rizzo, Magnetic and electric properties of quantum vacuum, Hal id: hal-00748532, https://hal.archives-ouvertes.fr/hal-00748532

24. S. Kocsis *et al.* Observing the average trajectories of single photons in a two-slit interferometer, Science. 2011 Jun 3;332(6034):1170-3. doi: 10.1126/science.1202218

25. D. Valev, Estimations of total mass and energy of the universe, arXiv:1004.1035v1 [physics.gen-ph] 7 Apr 2010

26. R. Battesti, C. Rizzo. Magnetic and electric properties of quantum vacuum. Reports on Progress in Physics, IOP Publishing, 2013, in press.

27. Yi Liang and Andrzej Czarnecki Alberta Thy 12-11 Photon-photon scattering: a tutorial arXiv:1111.6126v2 [hep-ph] 8 Dec 2011
Department of Physics, University of Alberta, Edmonton, Alberta, Canada T6G 2E1

Appendix I

Why Vacuum Polarization is not in Play

The first interaction between the Feynman photons and an interloping photon that comes to mind is the vacuum Polarization, which is at least twenty orders of magnitude greater than the effect of the proposed scattering and the effects of gravitation [22]. The interaction of the photon electromagnetic field, with static fields and other photons with strength enough to produce Electron-Positron pairs known as Vacuum Polarization is the most widely studied and developed Photon-Photon interaction.

Vacuum Polarization however is a vector effect, related to the electromagnetic field or as has been proposed, the probability density (Proposition 3). In the aggregate of a large number of random photon sources, the vector directions and phases cancel yielding an absence of any effect, and thus this effect is not likely to play a role in gravitation.

As a photon passes an oscillating photon path as defined above it is expected that there will be a probability of interaction between the passing photon and the Feynman photons. The most significant interaction for two interaction photons is the vacuum polarization defined by the Schwinger limit:

$$\frac{\Delta c}{c_0} = \frac{\bar{E}^2}{E_S^2} = \frac{\left(E^2 + B^2 + 2|S|\right)}{E_S^2} \tag{1}$$

The last term is the electromagnetic field density in terms of the electric magnetic, and Poynting vectors for a collision of two photons. E_S^2 is the Schwinger density and the fields are the sum of the electromagnetic vectors of the two photons[26].

From Y. Lang, et. al. [27] and R. Battesti, et. al. [23] the Vacuum Polarization photon cross section for photon-photon collision in the low energy approximation for unpolarized light, the total cross section can be written as:

$$\sigma_{\gamma\gamma\to\gamma\gamma} = \frac{973\alpha^2}{10125\pi} \frac{\lambdabar_e^6}{\lambdabar_{PH}^8} \tag{2}$$

This cross section is dependent of the mutual energy of the interaction photons thus any global effect of the speed of passing photons would be dispersive and inconsistent with a gravitational induced change in c. For a 1 ev photon this is about 1.34e-61 cm^2 for a electron mass equivalent photon this would be about 9.0e-40 cm^2

The source being considered, however are the Feynman photons of atoms, for which a mass source is a large number of randomly oriented positions and phases, and as the number of randomly oriented sources goes large, $(6.0 \times 10^{23}$ Avogadro's number), the sum of the vacuum polarization inducing electromagnetic vectors vanish.

$$E = \sum_n E_n \sin\theta_n \to 0$$
$$B = \sum_n E_n \cos\theta_n \to 0 \tag{3}$$

And the effects of vacuum polarization vanishes also vanish.

Vacuum polarization is a vector phenomenon, and just like an electric field can effectively cancel everywhere. The existence in space of the Feynman particles is a probability density that is conserved. Unlike Vacuum Polarization the accumulation of the probability density for massive particles is conserved regardless of the relative phases or size.

Appendix II

Cosmological c

From Eq.(13), the relation for the change in c in propinquity with a mass particle can be summed over all the particles in the observable universe to give the ambient value of c in the universe. From D. Valev [25], the value of this can be estimated to be:

$$M \approx \frac{c^3}{GH} \approx \frac{c^2 R}{G} \tag{1}$$

R is the radius of the universe, H is the Hubble constant and G is the Newton gravitation constant. This can be written as:

$$\sum \frac{mG}{c^2 R} = 1 \tag{2}$$

By presuming the average distance to any particle in the universe is about half the radius of the universe the value of c for the universe Eq.(13), can be found by summing over all the particles as:

$$\frac{\Delta c}{c_0} = \sum_n \frac{\lambda_{PL}^2}{\left(\lambda_{Ph}^2\right)_n} \frac{\left(2\lambda_{PH}\right)_n}{r_n} = n \frac{G}{c_0^2} \frac{M}{R/2} \approx 1 \tag{3}$$

M is the mass of the universe R is the radius and R/2 is on the order of the average distances r_n to each of the particles. Eq.(3), matches Eq(2). if $\Delta c \approx c_0$, indicating relation Eq.(13), which applies to the change in c induced by a single particle, also applies to the total of the mass particles in the universe and sets the ambient level of c.

Gravitation is a Gradient in the Velocity of Light

D.T. Froedge

V250117

Abstract

It is well known that a photon moving in a gravitational field has a trajectory that can be defined by Fermat's principle in Minkowski flat space with a variable speed of light and no other gravitational influence.

It can be shown that confined massless lightspeed sub-particles function inertially equivalent to a mass particle and have the same acceleration in a variable index of refraction as a mass particle in a gravitational field. If this is true then it is argued that the internal constituents of all mass depend on the velocity of light for internal propagation of its constituents and are accelerated in a gradient in c just as confined photons. The best evidence for this is that the energy change as the result of a Lorentz velocity transform is the same for particles and photons.

If mass particles are at the core, bound lightspeed particles then there is no need to ascribe any other mechanism to gravitation than a gradient in c.

This makes gravitation an electromagnetic phenomenon, and if QFT can illustrate a gradient in c equivalent to gravitation, can be produced by the internal motion of lightspeed sub-particles then the unification of QM and gravitation becomes more straightforward.*

http://www.arxdtf.org/css/GravAPS.pdf

*It has been asserted by researchers that the quantum vacuum is the origin of the speed of light and there is research on the phenomena, that the passage of a beam of photons can produce a gradient in c. [1][2]

255

Introduction

There are hundreds of papers on gravitation and a variable speed of light (VLS) some of which preexist General Relativity and many thereafter [3]. All papers so far reviewed by this author assert that gravitation alters the velocity of light as well as providing an attractive potential for mass particles.

It is assumed here and developed in "The Concept of Mass as Interfering Photons"[4] that a pair of bound of lightspeed particles having a mass as defined by $m = E/c$ have the same inertial properties and gravitational properties as a mass particle.

The effect of a gradient in c on such a particle functioning in a variable index of refraction following a trajectory defined by Fermat's principle is developed.

The initial horizontal trajectory of the back and forth motion of lightspeed particles accelerates vertically exactly as a mass particle in a gravitational field. As in previous papers regarding a locally conserved energy principle, flat Minkowski space is presumed. [5]

A simple particle

It is presumed that a simple particle such an electron could be defined bound lightspeed such as a photon, and a neutrino of equal energy having aligned spin and motion reciprocating initially along a common axis. The sum of the spin is ½ and the particles could be held together by a mediating W boson that instantaneously exchanges momentum when the particles exceed the combined particle Compton diameter.

The linear momentum of the light speed particles, developed in [6], can be defined by:

$$\vec{P} = \frac{h\nu}{c^2}\left(\gamma^k c_k + \gamma^0 c\right) \tag{1}$$

Summing, squaring two particles, and noting the value is Lorentz invariant gives:

$$\frac{(v_1+v_2)^2}{c^2}\left[1-\frac{(v_1-v_2)^2}{(v_1+v_2)^2}\right]=\frac{2v_1v_2}{c_0^2}=\frac{v_0^2}{c_0^2} \tag{2}$$

Noting that this expression is the same for the relativistic mass of a particle with $m=(v_1+v_2)/c^2$, and the velocity of the center of mass is:

$$\frac{v}{c}=\frac{(v_1-v_2)^2}{(v_1+v_2)^2} \tag{3}$$

and the deBroglie frequency is the difference of the photon frequencies.

The motion of the photons can be treated by Fermat's principle in a variable index of refraction. For illustrative purposes it is assumed that the particles are initially moving horizontally along the x axis with the index of refraction gradient along the vertical r axis with the spin being horizontal. As the lightspeed particles move back and forth in the confined volume the acceleration of the vertical component of the trajectories responds to the gradient in c as would a photon in free space.

The acceleration is actually independent of these initial conditions, and thus any configuration of a collection of confined lightspeed particles results in the combined particle center of mass mimicking a real particle in a gravitation field.

Gravitationally Induced Index of Refraction

The index of refraction of light induced by a gravitating mass can be deduced using the defection of starlight by the by the use of Fermat's principle; in addition Blandford & Thorne [7] have shown the same result by projecting the Einstein equations on flat space. That index of refraction is found to be:

$$\eta_\theta = \left(1 - 2\frac{\mu}{r}\right)^{-1} \tag{4}$$

This result, which is well verified for tangential or angular motion of photons in a gravitational field, is not necessarily the same as the radial value. In fact Karimi, & Khorasani [8], has illustrated that the Schwarzschild metric yields, for flat space, an index of refraction that is optically anisotropic.

Karimi, & Khorasani [6], have shown that with a more detailed development of the asymmetric aspects of the GR metric, that the index of refraction is actually:

$$\eta = (1+\phi)^{-1/2}\left(1+\phi\cos^2\theta\right)^{1/2} \quad , \quad \phi = \frac{r_s}{r} = \frac{2\mu}{r} \tag{5}$$

The angle θ is the angle between the wave vector and the radius. By dropping second order terms and simplifying, the velocity becomes:

$$\eta = \left(1 - \left(1+\cos^2\psi\right)\frac{\mu}{r}\right)^{-1} \tag{6}$$

θ is the angle between the velocity and the radius vector.

Horizontal Trajectory

The equations for photon movement in a variable index of refraction have long been worked out for lens development. From Evans & Rosenquist [10],[11] the equation for the trajectory of a photon in a variable index medium derived from variational principles based on Fermat's theorem is:

$$\mathbf{r}'' = \eta\nabla(\eta) \tag{7}$$

With the derivative defined with respect to a stepping parameter, such that:

$$da = \frac{c_0}{\eta^2} dt$$

(8)

And:

$$|\mathbf{r'}| = \left|\frac{d\mathbf{r}}{da}\right| = \eta$$

(9)

Inserting Eq.(8), into Eq.(7), and proceeding, the time differential or vertical acceleration is:

$$\frac{d^2r}{dt^2} = \left(\frac{Gm}{r^2}\right)\left[1 - \frac{v_r^2}{c_0^2}\right]^2$$

(10)

This is the same as the acceleration of a mass particle in gravitation for flat space.

(See Appendix I for math details.)

For the horizontal x component of the photon acceleration is:

$$\frac{d^2x}{dt^2} = \left(2\frac{Gm}{r^2}\right)\left[1 - \frac{v_x^2}{c_0^2}\right]^2 = 0$$

This is initially zero for a particle direction along the x axis. These equations can be solved for the trajectory and is illustrated in Figure 1.

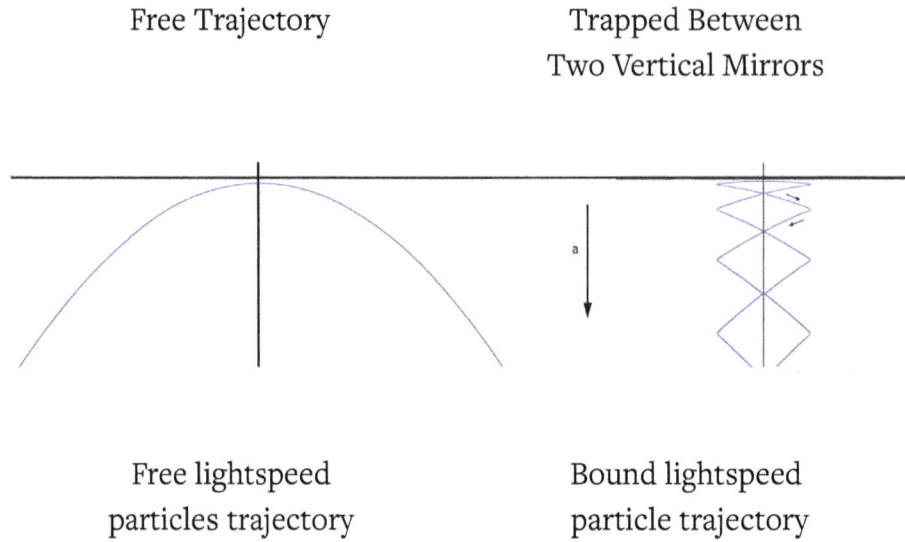

Free Trajectory	Trapped Between Two Vertical Mirrors
Free lightspeed particles trajectory	Bound lightspeed particle trajectory

The vertical acceleration for both bound photons traveling at light speed, and free particles in a variable index of refraction.

Fig. 1.

The initial conditions of a stationary center of mass and a horizontal trajectory are not necessary for the results.

Vertical Trajectory

The above illustrates the particles being initially aligned along the horizontal axis, but it is easy to show that the same result applies to any initial condition.

A more direct approach developed in earlier papers, [4], and easily shown, Eq.(3), the velocity of the center of mass, or "center of energy" for two photons, or speed of light particles, is:

$$\frac{\vec{v}}{c} = \frac{\left(v_1 \dfrac{\vec{c}_1}{c} + v_2 \dfrac{\vec{c}_2}{c} \right)}{\left(v_1 + v_2 \right)} \tag{11}$$

Presuming the particles initially of the same energy and are moving opposite directions along a vertical axis with a gradient in c, Fig 2, the acceleration of the center of mass of the two particles as the result of a gradient in c is:

260

$$\frac{dv}{dt} = \frac{d}{dt}\left[c\frac{\lambda_0}{2}\left(\frac{1}{\lambda_1} - \frac{1}{\lambda_2} \right) \right] \qquad (12)$$

With $\lambda_0 = \lambda_1 + \lambda_2$. Inserting the wavelength dependence on index of refraction from Eq. (6), foe a vertical trajectory, this becomes

$$\frac{dv}{dt} = \frac{c}{2}\left(\left(1 - \frac{\mu}{r_1} \right) - \left(1 - \frac{\mu}{r_2} \right) \right) \qquad (13)$$

Since the particles are moving in opposite directions $c = dr_1 / dt = -dr_2 / dt$, the vertical acceleration of the center of mass of the two lightspeed bound particles is:

$$a = \frac{dv}{dt} = \frac{Gm}{r^2} \qquad (14)$$

This is exactly as above Eq. (10), for the vertical acceleration of a bound system of two light speed particles, and the same for a massive particle moving in a gravitational field.

Particle Trajectories for Bound Particles Vertical Trajectories

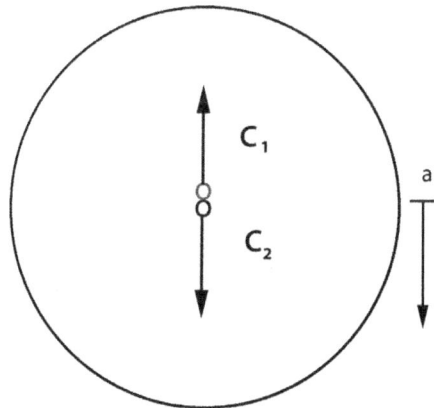

Figure 2

The Argument

1. A photon trapped between two reflectors whether mirrors or in a black box cavity represents rest mass. From the outside the photon is part of the total energy and cannot be ignored as part of the proper mass.
2. A photon reflecting between two parallel vertical mirrors accelerates vertically in a gradient in c exactly as a mass particle in a gravitational field.
3. If a trapped photon is mass then a photon oscillating between two mirrors must generate gravitation the same as a mass particle of the same energy.
4. An energetic photon can by virtue of QFT generate a gradient in c in its path vicinity.

Therefore; if the simplest form of mass, a trapped photon, can respond to a gradient in c as if it is in a gravitation field, and can generate a gradient in c equivalent to that produced by gravitation, why is there a necessity ascribe any other attribute to gravitation.

Conclusion

It has been shown that the effects of gravitation attraction of mass particles can be effectuated on a mass defined by a confinement of lightspeed particles by a gradient in the speed of light without need of any other mechanism. Since there are well-known processes defined in QFT and path integral formulations of QM, that alter the velocity of light in the proximity of moving photons,[1][2] it is speculated in appendix II, that the processes of QFT could be the progenitor of gravitation.

References:

1. M. Urban,et.al. The quantum vacuum as the origin of the speed of light, The European Physical Journal D 67(11):219- November 2013 http://arxiv.org/abs/1302.6165

2. D. Kharzeeva, K. Tuchinb, Vacuum Self–Focusing of Very Intense Laser Beams, arXiv:hep-ph/0611133v2

3 Jan Broekaert, 2008 A spatially-VSL gravity model with 1-PN limit of February 5mhttp://arxiv.org/pdf/gr-qc/0405015v4.pdf

4 DT Froedge, Gravitation is a Gradient in the Velocity of Light,,
V051715,
http://www.arxdtf.org/css/Gravitation.pdf

5. DT Froedge, The Concept of Mass as Interfering Photons,
V032615, http://www.arxdtf.org/css/interfering.pdf

6 DT Froedge, The Gravitational Theory with Local conservation of Energy, V020914, http://www.arxdtf.org/css/grav2a.pdf

7. Roger Blandford, Kip S. Thorne, Applications of Classical
Physics, (in preparation, 2004), Chapter 26
http://pmaweb.caltech.edu/Courses/ph136/yr2012/1227.1.K.pdf

8. F. Karimi, S. Khorasani, Ray-tracing and Interferometry in
Schwarzschild Geometry, arXiv:1001.2177
gr-qc] arXiv:1206.1947v1 [gr-qc] 9 Jun 2012

9. R.V. Pound and J.L. Snider, Effect of gravity on gamma radiation,
Phys. Rev. B
140:788–803 (1965).

10. James Evans and Mark Rosenquist, " 'F = ma' optics." *American Journal of Physics* 54 (1986) 876-883.
http://www2.ups.edu/faculty/jcevans/F=ma%20Optics.pdf

11. Simple forms for equations of rays in gradient-index lenses, James Evans *American Journal of Physics* 58 (1990)
http://www2.ups.edu/faculty/jcevans/Simple%20forms.pdf

12. DT Froedge, The Velocity of Light in a Locally Conserved Gravitational Field, V081216,)
http://www.arxdtf.org/css/velocity.pdf

Appendix I

Calculations

The presumption is that the initial motion of the lightspeed particle is tangential to the gradient in the velocity of light. With axis such that r is vertical and θ the velocity is orthogonal and along the x axis.

From above, the gravitational equivalent index of refraction for a speed of light particle along the , and directions are:

$$\eta_r = \lambda_0 \left(1 - 2\frac{\mu}{r}\right)^{-1/2} \qquad (1.1)$$

$$\eta_\theta = \left(1 - 2\frac{\mu}{r}\right)^{-1} \qquad (1.2)$$

From [10] the motion of a speed of light particle in a variable index of refraction is:

$$\mathbf{r}'' = \frac{d}{da}\frac{d\mathbf{r}}{da} = \text{grad}\left(\frac{\eta^2}{2}\right) \qquad (1.3)$$

With the stepping parameter a defined as:

$$da = \frac{c_0}{\eta^2}dt \qquad \text{and} \qquad |\mathbf{r}'| = \left|\frac{d\mathbf{r}}{da}\right| = n \qquad (1.4)$$

To express Eq.(1.3), in time dependent acceleration the stepping parameter can be replaced with Eq.(1.4), giving for the tangential motion along the r axis:

$$\frac{d}{da}\left(\frac{dr}{da}\right) = \frac{d}{da}\left(\frac{\eta^2}{c_0}\frac{dr}{dt}\right) = \frac{\eta^2}{c_0}\frac{d}{dt}\left(\frac{n^2}{c_0}\frac{dr}{dt}\right) = \frac{\eta^2}{c_0}\frac{d}{dt}\left(\frac{\eta^2}{c_0}\right) + \frac{\eta^2}{c_0}\frac{\eta^2}{c_0}\frac{d}{dt}\frac{dr}{dt} \qquad (1.5)$$

Thus:

$$\mathbf{a} = \frac{d}{dt}\frac{dr}{dt} = \frac{c_0^2}{\eta^4}\mathrm{grad}\left(\frac{\eta^2}{2}\right) - \frac{c_0}{\eta^2}\frac{d}{dt}\left(\frac{\eta^2}{c_0}\right) \qquad (1.6)$$

For the radial component of the acceleration is:

$$\frac{d^2r}{dt^2} = c_0^2\left(\frac{\mu}{r^2}\right)\left(1-\frac{\mu}{r}\right)^{-1} - v_r^2\left(\frac{2\mu}{r^2}\right)\left(1-\frac{\mu}{r}\right)^{1} \qquad (1.7)$$

or

$$\frac{d^2r}{dt^2} = \left(\frac{Gm}{r^2}\right)\left[1-\frac{v_r^2}{c_0^2}\right]^2 \qquad (1.8)$$

Similarly the horizontal x component of the photon acceleration is:

$$\mathbf{a_x} = \frac{d^2x}{dt^2} = \left(2\frac{Gm}{r^2}\right)\left[1-\frac{v_x^2}{c_0^2}\right]^2 = 0 \qquad (1.9)$$

The radial or vertical acceleration is exactly equivalent to the acceleration of a relativistic particle in flat space under the influence of a gravitational field.

These equations can be solved for the trajectory of the photon experiencing a gradient in c. See Evans [11] for the procedure.

Appendix II

It has been demonstrated above that the effect of gravitation can be mimicked on the dynamics of photons and particles by a gradient in c. It is certainly true for the dynamics of photons and confined photons. Given this; a demonstration that a photon reciprocating in a cavity generates a gradient in c, of the proper value, in the surrounding space by methods of QFT, would transform gravitation form a distinct force, to electromagnetic phenomena.

Since there are well-known processes defined in QFT and path integral formulations of QM, that alter the velocity of light in the proximity of moving particles,[10] ,[11] it is speculated that these processes could be the progenitor of the gravitational phenomena. It is a bit of speculation, but not farfetched with respect to known QFT phenomena.

It has long been known that a photon entering a gravitational potential follows a path identical to that of a photon in a variable speed of light defined by the Shapiro velocity for Minkowski flat space [7]. A spatially variable speed of light is implicitly present in General Relativity, and in fact has a long history starting in the pre GR efforts of Einstein and others. The difference in the approach taken in this author's paper is not that gravitation changes the speed of light, but that gravitation **is a change in the speed of light.** Newton's apple falls not because of an increase in energy, but because the speed of light at the branch is higher than the speed of light at the ground.

Simplifying the Discussion

First; it is observed that a photon confined in a reflective cavity constitutes rest mass. If to an empty cavity having a given mass is added a number of photons, the energy is increased, and for an outside observer the increase though small, constitutes an inertial or rest mass increase. For a black body cavity containing black body radiation the effect may be small, but in the case of an atomic nucleon the addition of a gamma ray is quite measurable. It must be concluded that the radiation confined in a black body cavity must contribute to the rest mass of the cavity and must be included in its inertial mass.

Second; if confined photons must be included in the rest mass, then the presence of the photons must also generate gravitational attraction. It must be concluded that a single photon bouncing back and forth between two reflectors is somehow generating the effect of gravitation, and in the view of this author a gradient in c.

Consider an apparatus having a cavity with opposing mirrors and having photons trapped between the mirrors. From conservation of energy the apparatus has more mass and generates more gravitational attraction than the cavity without the photons. There is not speculated an interaction between the photons, so the photons that are bouncing back and forth must be generating gravitation.

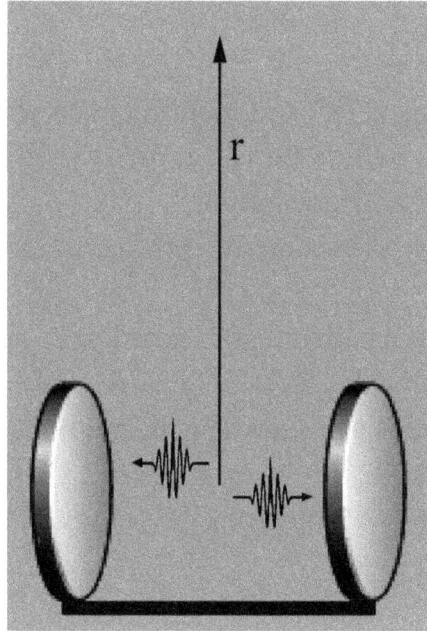

Fig1

Photons trapped between mirrors of an apparatus increase the mass and thus the gravitational attraction of the apparatus.

The increase in energy of the system is hv so the mass of the apparatus increase as a result of a trapped photon is:

$$m = \frac{\hbar\omega}{c^2} \tag{2.1}$$

The gravitational potential due to a confined photon is then:

$$\frac{\mu}{r} = \frac{G\hbar\omega}{c^4 r} \tag{2.2}$$

Putting this into the radial value for the index of refraction [13] of light in flat space yields:

$$c = c_0\left(1 - \frac{G\hbar\omega}{c^4 r}\right) \tag{2.3}$$

or:

$$\Delta c = \frac{G\hbar}{c^3 r}\omega \tag{2.4}$$

Noting that the square of the Planck radius is $G\hbar/c^3$ this can be stated as:

$$\Delta c = \frac{r_P^2}{r}\,\omega \qquad (2.5)$$

From our premise; "if" the motion of the photons generates a gradient in c, equivalent to Eq.(2.5), then the effect on other particles is equivalent that of gravitation.

The fact that the Planck radius is the constant in the equation is quite curious.

By the methods of path integrals noted by Feynman the probability for the particle moving from point a to point b, exist throughout spaces, it has already been shown by the methods of Quantum Electrodynamics that a photon beam induces a change in the velocity of light in the vicinity of the beam. [2], if this is the value then the conjecture will be proven.

This illustration shows the path actions induced by a photon oscillating between two reflectors.

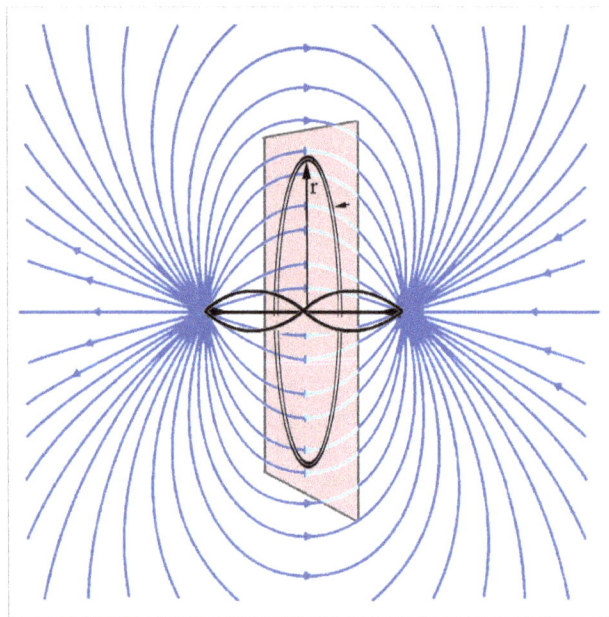

Fig.2.

From the work of D. Kharzeeva, et.al, [2] it is shown that for an intense laser beam the QFT effects related to electron–positron loops induce vacuum "self-focusing" which is a vacuum alteration of the index of refraction in the speed of light in the vicinity of the beam

A particle model being reciprocating bosons in a massless box, as asserted here, constitutes an intense, highly energetic back and forth reciprocal motion, orders of magnitude greater than a laser.

It is suggested here that the multiple path integration of the photon action over all space would alter the velocity of light near the path as a function of r, and if Eq. (2.5), is realized the connection between gravitation and QFT would be established.

* More succinctly the anisotropic gravitational induced velocity of light is, $c = c_0 \left(1 - (1 + \sin\theta)\mu / r\right)$ with θ being the angle between the velocity and the radius vector [12]

$$\eta = (1 + \phi)^{-1/2} \left(1 + \phi\cos^2\psi\right)^{1/2} \quad , \quad \phi = \frac{r_s}{r} = \frac{2\mu}{r} \tag{2.6}$$

The angle ψ ins the angle between the wave vector and the radius.

By dropping second order terms and simplifying, this becomes the same expression as was developed from experimental considerations in [8].

$$c = \left(1 - \left(1 + \cos^2\psi\right)\frac{\mu}{r}\right) \tag{2.7}$$

Due linearising of the gravitational potential in GR:

$$\left(1 - \frac{\mu}{r}\right)^2 \rightarrow \left(1 - 2\frac{\mu}{r}\right) \tag{2.8}$$

the Karimi expression Eq. (2.6), is not correct for a conservative field in which photons do not transfer energy to the field.

The propagation expression used in the original presentation was:

$$c = \left(1 - \frac{\mu}{r}\right)^2 \tag{2.9}$$

The expression for the asymmetric propagation of light near a gravitational radius used in the ray projections in this presentation is:

$$c = \left(1 - \frac{\mu}{r}\right)^{\left(1 + \cos^2\psi\right)} \tag{2.10}$$

A Physical Electron-Positron Model in Geometric Algebra

D.T. Froedge

V032417

Abstract

This paper will focus on a Geometric Algebra model of an electron illustrating the physical properties including the source of the charge. It will be shown that a particle such as an electron can be modeled as a pair of interaction photons bound in circular motion by their own interaction electric vectors, and having a Lorentz Scalar rest mass. This model is not QM, and does not have a probability interpretation, but rather it is defined in terms of an electromagnetic gauge. The notation used here is generally Feynman Slash [2] and the correspondence relations between GA and QM developed by Doran & Lasenby [3].

Given other options, there has been discussions by Gill, Hiley and others [4],[5]. as to whether Geometric Algebra is the best approach to defining the properties of quantum systems. A clear advantage, used here, is the ability in four-space geometry to design physical structures.

Note:
Technically the interacting speed of light constituents within a particle should not be referred to as photons, but it does seem to be the most useful mnemonic designation.

Systemfunction

The Systemfunction $\bar{\Theta}$ representing a particle is defined as a function that properly models both real and vector properties of a particle. The scalar factor is real and the vector is defined in terms of GA rotors [3].

Preliminary Constructs

The instantaneous null action vector for a photon can be defined as:

$$\mathcal{S} = m\left(\gamma^k c x_k + \gamma^0 c c t \right) \tag{1}$$

For convenience and clarity of concepts a mass m assigned to a photon and defined as $m = \hbar\omega / c^2$.

A Systemfunction for a photon is defined as:

$$\bar{\Theta}_F = e^{\left(\mathcal{S} + I_3 \right)\mathcal{S}} \tag{2}$$

Where I_3 is the grade three pseudoscalar $\gamma^1\gamma^2\gamma^3$

The photon Systemfunction is composed of both the standard action of QM and a scalar square of the action that functions as a Gaussian kernel defining the location of the particle moving along the light cone. It satisfies the physical as well as the vector properties of a photon.

Noting that $I_3\mathcal{S}$ is the QM action mapped into GA and that the magnitude of I_3 times the action is a solution of the Schrodinger equation.

Explicitly writing the function in GA gives:

$$\bar{\Theta}_F = e^{m^2_2\left(\gamma^k c x_k + \gamma^0 c c t \right)^2 + I_3 m\left(\gamma^k c x_k + \gamma^0 c c t \right)} \tag{3}$$

This function has all the features of a photon. The first term is a null Gaussian having a half-width of 1.4 λ, and moving at c. the second term defines a rotor in the σ^k plane perpendicular to the direction of travel with the frequency defined by $\hbar\omega = mc^2$, and $m = \hbar\omega / c^2$

Preliminaries

In the proposed model of the Electron and Positron, it will be shown that when the particles co-located at the same point the Systemfunction can be factored into a pair of opposite going photons i.e.:

$$\bar{\Theta} = \bar{\Theta}_E \bar{\Theta}_P = \bar{\Theta}_{F1} \bar{\Theta}_{F1} \tag{4}$$

As in earlier papers, the functional model for the particle is a physical model, and has been designated as the Systemfunction. The use of the Systemfunction designation avoids confusion with QM wavefunctions.

Momentum and Rest Mass

The four momentum of a particle is the instantaneous value of the derivative of the action of a particle, and thus a starting point for physical particle properties

It has been shown by the author that opposite going photons locked together exhibit dynamical propertied s of massive particles, [6] thus a starting point for dynamical particles.

For two photons going in opposite directions in Minkowski space, the momentum is:

$$\not{P}_1 = m_1 \left(\gamma^k c_{1k} + \gamma^0 c \right) \tag{5}$$

$$\not{P}_2 = m_2 \left(-\gamma^k c_{2k} + \gamma^0 c \right) \tag{6}$$

The sum is:

$$\not{P} = \not{P}_1 + \not{P}_2 = (m_1 + m_2)\gamma^0 c + (m_1 - m_2)\gamma^k c_{1k} \tag{7}$$

Squaring:

$$\left(\not{P}_1 + \not{P}_2 \right)^2 = 4\not{P}_1 \not{P}_2 = 4m_1 m_2 c^2 = m_0^2 c^2 \tag{8}$$

The fact that this is the rest mass comes from the product of two Lorentz invariant vectors is a Lorentz scalar constant, and thus invariant under a Lorentz transformation. [7]. This is the definition of a rest mass for two opposite photons. Note that there is a defined rest mass even if the photons are not physically in the same location.

Eq. (7), can also be written as:

$$\cancel{P} = (m_1 + m_2)\left(\gamma^0 c + \frac{(m_1 - m_2)}{(m_1 + m_2)}\gamma^k c_{1k}\right) \qquad (9)$$

For two such particles the velocity of the center of gravity v_C can be determined from:

$$(m_1 + m_2)v_C = m_1 c - m_2 c \qquad (10)$$

Thus the velocity of the center of mass of the two photons is related to their respective energy by:

$$\frac{v}{c} = \frac{(m_1 - m_2)}{(m_1 + m_2)} \qquad (11)$$

Squaring the square of the four momentum for the two photons becomes the standard invariant mass form:

$$\cancel{P}^2 = m^2 c^2 \left(1 - \frac{v^2}{c^2}\right) = m_0^2 c^2 \qquad (12)$$

This is the best indicator that two bound sped of light particles can be considered to have rest mass.

Action Four-Vector

The instantaneous four-derivative of the action must necessarily be the four vector momentum defined above Eq(6), thus:

$$\cancel{\partial}\cancel{S} = \cancel{P} \qquad (13)$$

where:

$$dS = m\left(\gamma^0 cdt + \gamma^k c_1 dx_k\right) \qquad (14)$$

274

The instantaneous particle action of the two photons Eq.(5), and Eq.(6), is then.

$$S_1 = m_1\left(\gamma^k c x_k + \gamma^0 c c t\right) \tag{15}$$

$$S_2 = m_2\left(-\gamma^k c x_k + \gamma^0 c c t\right) \tag{16}$$

And the sum of the instantaneous particle action of the two photons Eq.(9), is:

$$\mathcal{S} = \mathcal{S}_1 + \mathcal{S}_2 = m\left(\frac{v}{c}\gamma^k c_1 x_k + \gamma^0 c^2 t\right) \tag{17}$$

Proposed Structure of Electron

It is proposed that a pair of photons moving in opposite directions $\mathcal{S}_1$ and $\mathcal{S}_2$, as discussed above are not moving along linear trajectories , but are engaged in such a way that there is a separation between the trajectories, and are thus orbiting in the $\gamma^1\gamma^2$ plane.

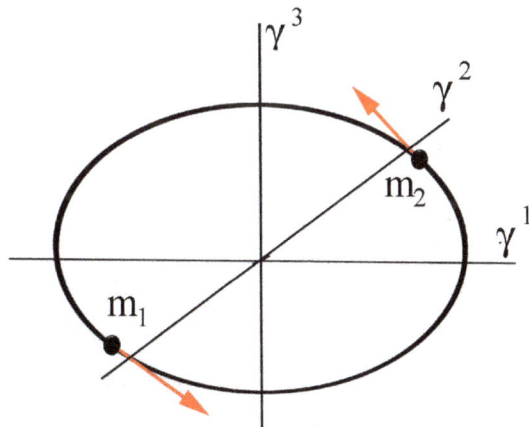

The Electron Systemfunction

It is proposed that the electron Systemfunction is composed of the action of two such opposite going photons Eq.(15), and Eq.(16), and is:

$$\bar{\Theta}_E = A e^{-\left[(\mathcal{S}_1 + \mathcal{S}_2) + I_3\right]^2 / 2} \tag{18}$$

Explicitly the exponents are:

$$-\left[(\mathcal{S}_1 + \mathcal{S}_2) + I_3\right]^2 / 2 = \frac{-\left[(\mathcal{S}_1^2 + \mathcal{S}_2^2) + (\mathcal{S}_1 \mathcal{S}_2 + \mathcal{S}_2 \mathcal{S}_1) + I_3^2 + I_3(\mathcal{S}_1 + \mathcal{S}_2) + (\mathcal{S}_1 + \mathcal{S}_2)I_3\right]}{2}$$

$$(19)$$

For convenience his can be separated into three functions the first a real scalar Gaussian, and two vector rotors.

The Gaussian:

$$\rho = \exp - \frac{\left(\mathcal{S}_1^2 + \mathcal{S}_2^2\right) + \left(\mathcal{S}_1 \mathcal{S}_2 + \mathcal{S}_2 \mathcal{S}_1\right) + 1}{2} \tag{20}$$

The first square terms are the null action vectors that define the Gaussian locations of the individual spin one photons.

The second bracket is the commutator of the photon action. Since the product of two Lorentz null vectors is a Lorentz constant this is a constant or twice the dot product of the opposite going particles. Using Eq.(8), this is just non kinetic rest mass, and for spin one photons the value of this term is:

$$\frac{(S_1 S_2 + S_2 S_1)}{2} = \frac{2 S_2 \cdot S_1}{2} = m_1 m_2 c^2 \left[\lambda_1^2 + (cT)^2\right] = 2 m_1 m_2 c^2 \lambda_1 \lambda_2 \tag{21}$$

From Eq.(8), we can then find that the value of the scalar simplifies to:

$$\rho = \exp-\left(\frac{{\not{S}_1}^2 + {\not{S}_2}^2}{2} + \frac{m_0^2 c^2 \lambda_1^2}{2} + \frac{1}{2} \right) \qquad (22)$$

From Eq.(19), the two defined rotors are:

$$R_{E1} = \exp\left[I_3 \left(\not{S}_1 + \not{S}_2 \right)/2 \right] \; = \; \exp\left[I_3 \not{S}/2 \right] \qquad (23)$$

and:

$$R_{E2} = \exp\left[\left(\not{S}_1 + \not{S}_2 \right) I_3 /2 \right] \; = \; \exp\left[\not{S} I_3 /2 \right] \qquad (24)$$

The product of these two rotors is not a null and thus the term is not Lorentz a null invariant.

$$\bar{\Theta}_E = \rho R_{E1} R_{E2} \qquad (25)$$

That is because R_{E1} and R_{E2}, are time reversals, and the product has only the three space vector momentum, not the time energy terms.

This is addressed by adding the initial condition of a π phase shift between $\not{S}_1$ and $\not{S}_2$, with that Eq.(25), then becomes:

$$\bar{\Theta}_E = \exp-\left[\left({\not{S}_1}^2 + {\not{S}_2}^2 \right) I_3/2 \; + \; I_3\left(\not{S}_1 + \not{S}_2 \right)/2 + \left(\not{S}_1 + \not{S}_2 \right) I_3 /2 + I\sigma^3 \frac{\pi}{2} \right]$$
$$(26)$$

Which can be written [3].as

$$\bar{\Theta}_E = \left(\rho^{1/2} R_{E1} \right) I_3 \sigma^3 \left(\rho^{1/2} R_{E2} \right) \; = \; \left(\rho^{1/2} R_{E1} \right)\left(\rho^{1/2} \tilde{R}_{E2} \right) I_3 \sigma^3$$
$$(27)$$

This is Lorentz invariant function. The physical meaning of the phase shift will become apparent in the physical description of the function

Electric Charge

For the electric charge we turn to the vector terms in the action and note that:

$$R_{E1} = e^{\left[I_3\left(\mathcal{S}_1 + \mathcal{S}_2\right)/2\right]}$$
(28)

or

$$R_{E1} = e^{\left[I_3\gamma^3\left(\omega_1 t + \omega_1 t\right)\right]/2}$$
(29)

Adding in the mechanical action of the particles revolving around the center of mass gives:

$$I_3\gamma^3\left(\omega_1 t + \omega_2 t + \left(\frac{m_1 c^2}{n\hbar}\right)t + \left(\frac{m_2 c^2}{n\hbar}t\right)\right)$$
(30)

Where n is the number of wavelengths of the photons from the center of mass:

$$\rightarrow I_3\gamma^3\left(\omega_1 t + \left(\omega_2 t + \pi\right) + \left(\frac{m_1 c^2}{n\hbar}\right)t + \left(\frac{m_2 c^2 t}{n\hbar}\right)\right)$$
(31)

or:

$$\rightarrow I_3\gamma^3\left(\left(\omega_1 t - \omega_2 t\right) + \left(\frac{\left(m_1 + m_2 c\right)}{n\hbar}\right)ct\right)$$
(32)

Or

$$\rightarrow I_3\gamma^3\left(\omega_1 t - \omega_2 t + \omega_R t\right)$$
(33)

From the model Fig 1. the two photons are revolving about their center of mass, and noting from Eq.(15), and (16) the photon internal rotation vectors are anti-aligned but in phase with the rotation phase, thus as the particles rotate around the center of mass the electric vector is always in a positive radial direction ie:

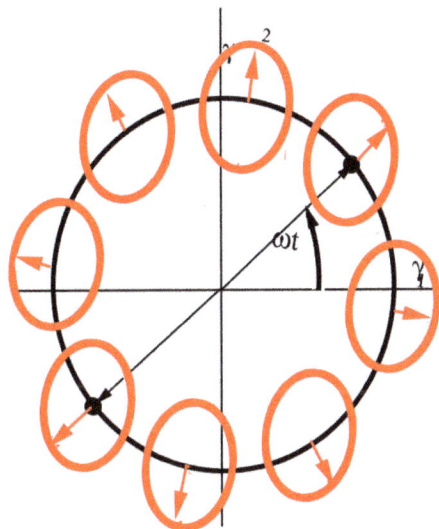

A first thought is that the vectors cancel and have are null, but since they are not co-located there is a Gaussian spatial relation. The vectors are only null at the center of mass.

Two photons orbiting in phase with the rotation can be either in phase or out with a phase difference of π in the rotational angle.

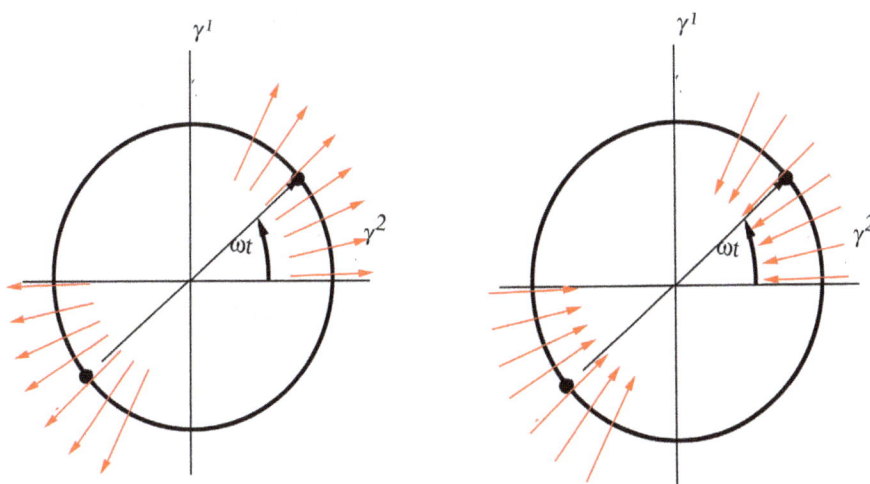

We can designate one of these as having the charge of an Electron and the other as a Positron.

These last configurations have electric vectors continuously positive or negative in the radial direction and a value of zero at the center of mass. The effect of one on the other is as two opposite charges and thus providing a radial binding force between the photons.

The discussion above only includes the electric vectors, but including the perpendicular magnetic vectors is straightforward and well described in [3].

Positron Systemfunction

The Positron Systemfunction is nearly identical to the electron accept for a space time reversal. From the definition of the Electron Systemfunction Eq. (18), a simple reflection of the spacetime of one of the factors gives the positron to be:

$$\bar{\Theta}_E = A e^{-\left[\left(\mathcal{S}_1 + \mathcal{S}_2\right) + I_3\right]\left[\left(\mathcal{S}_1 - \mathcal{S}_2\right) - I_3\right]/2} \qquad (34)$$

Note that the negative signs just reverse the time & space of $\mathcal{S}_2$ and I_3 otherwise Eq. (18), and Eq. (34), are identical. The photon designations for the Positron of 1 and 2 are kept the same since they are equivalent.

Displaying the functions together illustrates the common terms.

$$\bar{\Theta}_E = \exp\frac{-\left[\left(\mathcal{S}_1^2 + \mathcal{S}_2^2\right) + \left(\mathcal{S}_1\mathcal{S}_2 + \mathcal{S}_2\mathcal{S}_1\right) + 1 + I_3\left(\mathcal{S}_1 + \mathcal{S}_2\right) + \left(\mathcal{S}_1 + \mathcal{S}_2\right)I_3 + I\sigma^3\frac{\pi}{2}\right]}{2}$$

$$\bar{\Theta}_P = \exp\frac{-\left[\left(\mathcal{S}_1^2 + \mathcal{S}_2^2\right) - \left(\mathcal{S}_1\mathcal{S}_2 + \mathcal{S}_2\mathcal{S}_1\right) - 1 + I_3\left(\mathcal{S}_1 - \mathcal{S}_2\right) + \left(\mathcal{S}_1 - \mathcal{S}_2\right)I_3 - I\sigma^3\frac{\pi}{2}\right]}{2}$$

$$(35)$$

For clarity: The first bracketed terms are the null light speed photons, the second is a rest mass associated with the binding, the product of I_3 with itself or its reverse designates the positive or negative mass, the third and fourth terms are the electric rotation vectors and the last are the phase relation between the photons

Noting that:

$$I_3 \cancel{s}_2 - \cancel{s}_2 I_3 = 0 \tag{36}$$

The product an electron and a photon located at the same point can be reduced to:

$$\bar{\Theta}_E \bar{\Theta}_P = \exp{-\left(\cancel{s}_1^{\,2} + \cancel{s}_2^{\,2} + I_3 \cancel{s}_1 + \cancel{s}_2 I_3 \right)} \tag{37}$$

This re-factoring shows the function to be the collocation of two opposite going spin-one photons;

$$\bar{\Theta}_{F1} \bar{\Theta}_{F2} = \exp{\left(\cancel{s}_1^{\,2} + I_3 \cancel{s}_1 \right)} \times \exp{\left(\cancel{s}_2^{\,2} + \cancel{s}_2 I_3 \right)} \tag{38}$$

This is the product of the Systemfunction of two photons defined in Eq.(2), having separate Gaussian kernels and no rest mass,

This is the clear indication that the electron positron pair or the opposite photon pair are equivalent and illustrate the annihilation process

Rest Mass and Binding Force

The invariant constant terms of the photon commutator can be recognized as a binding energy.

$$\left(\cancel{s}_1 \cancel{s}_2 + \cancel{s}_2 \cancel{s}_1 \right) \tag{39}$$

As the photons rotate the electric vectors are opposite and create an attractive binding force between the opposite photon. Since the force must be equivalent to the centrifugal force the magnitude of this force can be evaluated.

$$F_c = \frac{m_1 v^2}{r} = \frac{m_1 c^2}{\lambda_1} = \frac{m_1 c}{\lambda_1 \hbar} c\hbar = \frac{c\hbar}{\lambda_1^{\,2}} = \frac{1}{\alpha} \frac{Q^2}{r} \tag{40}$$

This is equivalent to the force on a charge at the classical electron radius, or 137 times greater than a charge-charge binding force at the rotation position. It should not be a surprise that the external electric vector is a factor of α less than the electric vector binding the photons.

Corresponding Dirac KG
Differential Solutions

It is relatively easy to show that the Electron Systemfunction is a solution to the equivalent of the Dirac equation.

At the center of mass, the square of the action is zero or at least constant and thus ρ can be considered constant. The entire System-function and factors are:

$$\bar{\Theta}_E = \rho R_{E1} I \sigma^3 R_{E2} \tag{41}$$

$$R_{E1} = \exp\left[I_3 \mathcal{S} / 2\right] \tag{42}$$

and:

$$R_{E2} = \exp\left[\mathcal{S} I_3 / 2\right] \tag{43}$$

The GA vector solution of the Dirac expression is then:

$$\mathcal{\partial}\, \bar{\Theta} = \mathcal{\partial}\, \rho R_{E1}\, I_3 \sigma^3\, R_{E2} = P\, \bar{\Theta} \tag{44}$$

The magnitude of which is the rest mass.

Note that the sum of the time inverted rotors differentials is:

$$\mathcal{\partial}\left(I_3 R_{E1} + R_{E2} I_3\right) I = P\left(R_{E1} + R_{E2}\right) \tag{45}$$

The second derivative of the Systemfunction is:

$$\mathcal{\partial}\mathcal{\partial}\, \bar{\Theta} = \mathcal{\partial}\mathcal{\partial}\, R_{E1} I \sigma^3 R_{E2} = P g P\, \bar{\Theta} = -m_0^2\, \bar{\Theta} \tag{46}$$

The function thus satisfies the GA equivalent of the Dirac [8] and the Klein Gordon expression.

Spin

The spin of the particles is just the time derivative of the rotor or:

$$\frac{d}{dt} R_{E1} = \frac{d}{dt} \exp\left[I_3 \mathcal{S} / 2\right] = \frac{1}{2} h R_{E1} \sigma^3 \tag{47}$$

This is identifiable as the proper spin for the electron moving along the γ^3 axis.

Binding Energy Annihilation and the Connection to QFT

The photon action as asserted in Eq.(1), could be termed as the classical value for a free particle. That is the path that the classical particle takes. From QFT, it is in fact the summation of all the possible paths that the particle could take. The QFT version of this path integral is found from [7]:

$$\psi\left(x_f, t_f\right) = \int dx_i \left[\int Dx(t) e^{i \int \frac{L}{\hbar} dt} \right] \psi\left(x_i, t_i\right) = \int dx_i \left[\int Dx(t) e^{i \frac{S}{\hbar}} \right] \psi\left(x_i, t_i\right)$$

(48)

If the particle is in the presence of a second particle the action must include not only the momentum but also the potential interaction of that particle with others. It is clear from the proposed model of the electron that the commutator terms of the action constitute the non-inertial binding energy of the Lagrangian holding the particle together.

It is beyond the scope of this paper to actually calculate the interaction from Path integral formulations, but since the interaction of a charged particle in the presence of an electron is well known, it is possible to include the effect into the electron model.

From Eq.(14), the vector photon action is:

$$dS = m\left(\gamma^0 cdt + \gamma^k c_1 dx_k\right)$$

(49)

and the non-dynamic energy term γ^0 term can be written as:

$$dS = \frac{m_1 c}{\hbar} d(ct) \rightarrow \frac{m_1 c}{\hbar} d(ct) \rightarrow \frac{\hbar c}{\lambda_1} \frac{dt}{\hbar} \rightarrow \frac{Q^2}{\alpha \lambda_1} \frac{dt}{\hbar}$$

(50)

From earlier Eq.(40), it was shown that $Q^2 / \alpha \lambda_1^2$ is equivalent to the binding force between the photons thus the $Q^2 / \alpha \lambda_1$, is be presumed to be a binding energy of the two photons.

For the Lagrangian of the electron in the presence of a second charged particle the energy associated with that interaction can be added. The added term from the second particle would have to be:

$$\int_9^{\lambdabar} E \frac{dt}{\hbar} = \int_9^{\lambdabar} \frac{Q^2}{r} \frac{dt}{\hbar} = \int_9^{\lambdabar} \frac{Q^2}{n\lambda_1} \frac{dt}{\hbar} = \int_9^{\lambdabar} \frac{\alpha}{n} \frac{dct}{\lambda_1} \qquad (1.51)$$

The distance to the exterior particle has been substituted with $n\lambda_1$, the number of wavelengths of the photon to the particle.

The integral of the Lagrangian is then:

$$\int_9^{\lambdabar} \left(\frac{\hbar c}{\lambda_1} \frac{dt}{\hbar} + \frac{\alpha}{n} \frac{dct}{\lambda_1} \right) \qquad (52)$$

Since the wavelength is $\lambda = 2\pi\lambdabar$, this becomes s modification to the photon four action γ^0 of:

$$\gamma^0 S \rightarrow \frac{ct}{\lambda_1} \left(1 + \frac{\alpha}{2\pi n} \right) \qquad (53)$$

If the exterior electron is in circular orbit then the action is the integral around its orbit. The term becomes the spin orbit coupling term and the second term in the spin g orbit coupling factor. The electron spin from Eq. (47), is then:

$$\frac{d}{dt} \exp\left[I_3 \mathcal{S} / 2 \right] = \frac{1}{2} h \left(1 + \frac{\alpha}{2\pi} \right) R_{E1} \sigma^3 \qquad (54)$$

This is not a derivation of the anomalous spin g-factor, but an illustration its mechanical connection to the model.

Note that in Eq. (52), the sign of the exterior particle can be opposite the electron, and if the distance to the exterior particle is reduced to the classical electron radius $r \rightarrow \alpha\lambda_1$ the rest mass terms of the commentators vanishes.

$$\frac{Q^2}{r} \frac{dt}{\hbar} \rightarrow -\frac{\hbar c}{\lambda_1} \frac{dt}{\hbar} \qquad (55)$$

This is the point of annihilation as shown in Eq. (35), & Eq. (38), at which point the photons are free to escape.

Neutrino

To illustrate the utility of the GA functional model of the electron, a plausible model for a neutrino can be constructed. Starting with the same function defining the electron, Eq.(18):

$$\bar{\Theta}_N = A e^{-\left[\left(\not{S}_1 + \not{S}_2\right) + I_3\right]^2} \tag{56}$$

but setting the action of the second photon $\not{S}_2$, to be in the same direction as $\not{S}_1$ but with an opposite phase, and an energy half the energy of $\not{S}_1$. Expanding this Eq.(18), gives:

$$\bar{\Theta}_N = \exp\left[\left(\not{S}_1 + \not{S}_2\right) + I_3\right]^2 / 2 = \exp-\left[\begin{array}{c}\left(\not{S}_1^2 + \not{S}_2^2\right) + 1 + I_3\sigma^3\dfrac{\pi}{2} \\ + I_3\left(\not{S}_1 + \not{S}_2\right) + \left(\not{S}_1 + \not{S}_2\right)I_3\end{array}\right] \tag{57}$$

With an opposite phase $m_2 = m_1 / 2$ and the Euler terms including a $\pi / 2$ phase shift then are:

$$I_3\left[m_1\left(\gamma^k c_1 x_k + \gamma^0 c^2 t\right) - \frac{1}{2}m_1\left(\gamma^k c_1 x_k + \gamma^0 c^2 t\right)\right]$$
$$+ \left[m_1\left(\gamma^k c_1 x_k + \gamma^0 c^2 t\right) - \frac{1}{2}m_1\left(\gamma^k c_1 x_k + \gamma^0 c^2 t\right)\right]I_3 + I_3\sigma^3\pi / 2 \tag{58}$$

Integrated the actions together gives:

$$I_3\left[\left(m_1 - \frac{1}{2}m_1\right)\oint\left(\gamma^k c_1 dx_k + \gamma^0 c^2 dt\right)\right] \tag{59}$$

$$I_3\left[\frac{S_1}{2}\right] + \left[\frac{S_1}{2}\right]I_3 + I_3\sigma^3 \tag{60}$$

The Systemfunction for the neutrino would then be:

$$\bar{\Theta}_{N} = \exp-\left[\left(\cancel{\delta}_{1}^{2} + \cancel{\delta}_{2}^{2}\right) + I_{3}\frac{\cancel{\delta}}{2} + \frac{\cancel{\delta}}{2}I_{3} + 1 + I_{3}\sigma^{3}\frac{\pi}{2}\right] \qquad (61)$$

This is two light speed photons traveling together, having an spin action that is half the spin of a single photon.

To examine the physical relation between the two particles, the function can be evaluated at the time and location of particle 1, which is t, $x_{1} = 0$. At that point if the second particle is located at a distance along the same trajectory at a fixed distance defined by:

$$\left(\frac{x_{2}}{\lambda_{2}}\right)^{2} - 1 = 0 \qquad (62)$$

There is only a single Gaussian node.

Similarly if the evaluation is made at particle 2, it is found that if particle 1 is at a fixed distance of λ_{1}, there is just a single node..

This is then two separate modes that give the appearance of a single particle having a spin of 1/2, This is suggestive of a particle with two modes representing different flavors.

As with the Positron Eq. (34), shown above, the product of the pseudoscalar could be with a reverse giving a negative neutrino with a -1, and negative half spin.

Conclusion

This paper presents a plausible complete GA model of photons, electrons, positrons, and the electron-positron pair annihilation process. Also presented is a plausible neutrino model, as well as a structure for other particles. It gives a more mechanical view of QM, and with the incorporation of QFT could lead to a solution of the mass ratio problem.

References:

1. D.T. Froedge, Particle Solution to the Klein-Gordon-Dirac Equation in the Context of a Big Bang Universe V031616, http://www.arxdtf.org

2. Weinberg, Steven (1995), The Quantum Theory of Fields, 1, Cambridge University Press, p. 358, ISBN 0-521-55001-7

3. C. Doran, A. Lasenby 2003 Geometric Algebra for Physicists, Cambridge University Press,

4. Richard D. Gill Does Geometric Algebra provides loop holes Bell's Theorem? 18 May, 2015Mathematical Institute, University of Leiden, Netherlands
http://vixra.org/abs/1504.0102 http://www.math.leidenuniv.nl/~gill

5. J. Hiley and R. E. Callaghan The Clifford Algebra approach to Quantum Mechanics A: The Schrodinger and Pauli Particles. University of London,
https://arxiv.org/abs/1011.4031

6. DT Froedge, The Concept of Mass as Interfering Photons, V032615 http://www.arxdtf.org/

5. Steven Errede The Structure of Space-Time Physics Lect. Notes 16 436 EM Fields & Sources II Fall Semester, 2015Department of Physics, University of Illinois at Urbana-Champaign, Illinois

7. Feynman, Hibbs, 1965, Quantum Mechanics and Path Integrals McGraw-Hill

8. P. A. M. Dirac, The Principles of Quantum Mechanics, 4th ed., Oxford University Press,London (1958).

Particle Solution to the Klein-Gordon-Dirac Equation in the Context of a Big Bang Universe

D.T. Froedge

V031616

Abstract

The purpose of this paper is to develop a single solution to both the Klein-Gordon & Dirac equations that expresses both the QM and the classical aspects of particles. It is found that this can be done, but only in the context of a system that has an initial event (T = 0) and is expanding at c, thus it is consistent with a big bang representation of the universe. The equations are defined in *geometric algebraic*, and the KGD equation will be considered a single equation factorable into linear products of the two linear Dirac expressions, with a solution defined analogously to path integrals. The solution has botha Gaussian shaped amplitude, (classical), and phase, (QM) components satisfying the quadratic KG equation, and the linear Dirac expression. The equation differentials are not restricted to representing the normal QM operator replacement of p and E, applicable to the linear equation, but have a broader context in operating on the more complex function with amplitude and phase factors. The solutions represent the particle at a single event, thus the standard view of the solution being a probability amplitude field over spacetime is not applicable, but an alternate observational field is illustrated that demonstrates the connection of the solutions to the observed wave characteristics. The phase factors are as usual cyclic, but the amplitude factors exist only in the context of the entire interval. The amplitude factor of the solution is proportional to mass and thus should offer insight into particle mass ratios.

Keywords: Klein-Gordon-Dirac, particles

Introduction

The purpose of this work is to develop a single point particle solution the KGD that defines the phase, quantum mechanical as well as the classical particle-particle dynamics and the electromagnetic interactions of particles. Instead of regarding the Klein-Gordon, and Dirac equations as analogs of the classical equations with the momentum and energy replaced by differential operators, the equations are treated as differentials operating on complex particle function.

$$\not{p} = m\not{\psi} \neq ih\not{\partial}\psi \tag{1}$$

The coordinates in this development are the end event coordinates of a particle solution, and not field variables of the function.

The particle solution will be formulated as the end point of a propagator, with function exponents that are the square of the sum of the action of the canonical momentum, over a classical fourspace interval from the initial, (Big Bang), to the current event. The end point function will be designated as the Systemfunction.

Section I, forms an analogy with the path integral approach, and proposes a particle solution that is the propagator, with an action that is the square of the fourspace bivector action of a particle from the Big Bang to its current position. The square of the complex bivector has both first and second order terms which are both real and imaginary. The gauge field is evaluated at the action endpoint making the solution a point function, and an eigensolution for the particle.

Section II separates a point Systemfunction into amplitude and phase functions. The complete Systemfunction is a solution to the KG expression, and also a solution to the Dirac expression.

Section III Shows that the phase factor of the point Systemfunction is equivalent to the Dirac solution for the first order equation, and develops the function properties.

Section IV defines an observational field associated with a point particle, having features similar to the Dirac probability amplitude field. The observational field illustrates planar deBroglie waves, and spherical Compton waves with a phase velocity of c.

Section V illustrates the point Systemfunction is a solutions to the classical KG equation, showing the relativistic mechanical properties, and the classical electromagnetic particle-particle interactions.

Preliminaries

A. Reviewing the standard QM relations

Free KG:

$$\left(-\frac{\partial^2}{\partial x^2} - \frac{\partial^2}{\partial y^2} - \frac{\partial^2}{\partial z^2} + \frac{\partial^2}{\partial t^2}\right)\psi = -m_0^2\psi \tag{2}$$

or in Feynman slash notation:

$$\left(\slashed{\partial}^2 + m_0^2\right)\psi = 0 \qquad\qquad \slashed{\partial} = \gamma^\mu \partial_\mu \tag{3}$$

The mass m_0 is the invariant rest mass, and ψ is the one component KG field.

Dirac:

$$\left(i\slashed{\partial} - m_0\right)\psi = 0 \tag{4}$$

Dirac with potential operator:

$$\left(i\slashed{\partial} - Q\slashed{A} - m_0\right)\psi = 0 \tag{5}$$

The wavefunction, ψ in this case is the Dirac spinor field, and $\slashed{A}$ is the electromagnetic gauge field in which the charge is immersed.

Schrödinger free particle:

$$\left(h^2\partial^k\partial_k + 2im_0\partial_0\right)\psi = 0 \tag{6}$$

The notation is such that the rest mass is the reciprocal of the Compton radius $1/D_0 = m_0 c/h \to m_0$, and he units are the *natural units* $h = c = 1$, except for clarification at section ends. (For general conventions, and notation, see **Appendix I**)

The point solutions in this paper will be designated as single component scalar, *Systemfunction* $\tilde{\Theta}$, to distinguish from standard wavefunctions, and it is presumed to be a solution to:

$$\left(\partial^2 + m_0^2 \right) \tilde{\Theta} = 0 \tag{7}$$

If the Systemfunction is factored into a real amplitude $\left(\tilde{\Theta}_R \right)$ and phase $\left(\tilde{\Theta}_I \right)$ functions such that:

$$\tilde{\Theta} = \tilde{\Theta}_R \tilde{\Theta}_I \tag{8}$$

Separation

Using the standard geometrical algebra factoring of the KG operator using the Clifford-Dirac gamma matrix Eq. (7), is:

$$\left(i\partial + m_0 \right) \left(i\partial - m_0 \right) \tilde{\Theta} = 0 \tag{9}$$

It is easy to show by the chain rule that $\tilde{\Theta}_I$ cannot be a solution of:

$$\left(i\partial - m_0 \right) \tilde{\Theta}_I \neq 0 \tag{10}$$

Unless any amplitude factor is a solution of:

$$\partial \tilde{\Theta}_R = 0 \tag{11}$$

See **Note 2.**

This implies that the gradient of the amplitude function is a constant, or is a null vector, thus having a phase velocity of c. This restriction on $\tilde{\Theta}_R$ can be met in the context of coordinate system originating at the initial event at $x = 0$, and $t = 0$, and traveling away from that event at the speed of c (see later development).

The linear expression for the phase, excluding the amplitude, is obtained directly from Eq(9). with the condition of Eq.(11):

$$\left(i\not{\partial} + m_0\right)\left[\left(i\not{\partial} - m_0\right)\tilde{\Theta}_I\right]\tilde{\Theta}_R = 0 \tag{12}$$

Since, Det $|AB|$ = Det $|A|$ × Det $|B|$ the square brackets can be picked from Eq.(12), and set as:

$$\text{Det}\left|\left(i\not{\partial} - m_0\right)\tilde{\Theta}_I\right| = 0 \tag{13}$$

The eigenvector equation for this expression is thus:

$$i\not{\partial}\, \mathbf{a}\tilde{\Theta}_I = m_0\, \mathbf{a}\tilde{\Theta}_I \tag{14}$$

The complete Systemfunction is a solution to equation Eq.(7), the amplitude is a solution to Eq.(11), and the phase part of the Systemfunction is a solution to Eq.(14),

The factorization has also induced an expansion of the number of solutions to the equation Eq.(7), by four. These extra solutions are for spin ½ particles, and the expression is identical in form to the Dirac expression if $\psi = \mathbf{a}\tilde{\Theta}_I$

B. Action & Wavefunctions

In this section a particle based solution, to Eq.(7), and Eq.(14), will be developed based of the square of a classical path of action, over a Minkowski fourspace interval defined over the life of the system of particles.

In the Path Integral formulation of QM, the amplitude of the probability for the m particle to transition from one state to another is the integral over a scalar action between the two states over all possible paths. The path integral depends on the final coordinate and time in such a way that it obeys the Schrödinger equation, [2], thus heuristically for a Schrödinger wavefunction:

$$\psi(x_f, t_f) = \int dx_i \left[\int Dx(t) e^{i \int \frac{L}{\hbar} dt} \right] \psi(x_i, t_i) = \int dx_i \left[\int Dx(t) e^{i \frac{S}{\hbar}} \right] \psi(x_i, t_i) \quad (15)$$

Where the propagator for the wavefunction going from an initial state i to a final state f is:

$$K(z_f t_f; x_i t_i) = \int Dx(t) e^{i \frac{S}{\hbar}} \quad (16)$$

And S is the particle action.

Analogously this development will propose a point solution to the expressions for Eq.(7), and Eq.(14), that represents a single particle, starting in the system at the Big Bang, (T = 0) and transitioning to the current event.

Observation of the wavefunction for a system of particles in the Schrodinger picture from Eq.(15), suggests that for a system of particles coming into existence the initial state at the big bang should be a constant $(x_i, T_i = 0, 0)$.

$$\psi(x_{1i}, t_{1i} \cdots x_{ni}, t_{ni}) = \text{constant} \quad (17)$$

Thus the primary constituent of the wavefunction going from initial to the current state is just the propagator. The path for the propagator considered here will be the classical path which is sufficient to illustrate the concepts. Path integral summation would induce refinements, but as in standard methodology the general picture would not be changed.

The Systemfunction would thus become:

$$\Theta = \int dx_i \left[\int Dx(t) e^{i \frac{S}{\hbar}} \right] \psi(x_i, t_i) \rightarrow A e^{i \frac{S}{\hbar}} \quad (18)$$

C. Defining the Lagrangian

The action of Eq.(16), and, Eq.(18), for a particle is in general the time integral of the path, and in the classical view the path taken minimizes this functional.

For the relativistic action of a particle it is presumed that a particle follows a path through 4-space that minimizes the space-time interval. In curved space-time, this path would be geodesic, but for our purposes, spacetime will be considered as flat, and the Minkowski action is:

$$S = \int_{t_i}^{t_f} L dt d^3 x \qquad (19)$$

L is an invariant, t & x are the spacetime coordinates, and S is an invariant action integrated over the spacetime interval. (Henceforth, h = 1).

For defining the Lagrangian, the spacetime vector four potential generated by a charged particle is.

$$\slashed{A} = \pm \alpha \left(\frac{\slashed{v}}{r} \right) \qquad (20)$$

Where $\slashed{v}$ the four-velocity, r is the distance to the particle, ± is the charge of the particle, and α is the fine structure constant [1]. (This is slightly a nonstandard notation, since it contains Q^2, but it is helpful this development.)

From experimental observations of charged particles, there is a photon scattering sphere at the classical electron radius, αD which is also the radius at which the integral of the electric energy equals the mass.

For a charged particle, it is proposed that the potential is actually a function of the distance to the charge radius and not the central point. Instead of the potential prescribed by Eq.(20), the gauge potential for charged particles is proposed to be:

$$\slashed{A} = \pm \alpha \left(\frac{\slashed{v}}{r - \alpha D} \right) \qquad (21)$$

The $\pm_j$ is the charge sign of the particle, and $D_j = h/mc$, is the Compton radius.

Since the photon cross section of a particle decreases as the velocity is increased [8], the mass in D is the relativistic mass, and the function is shown in figure 1.

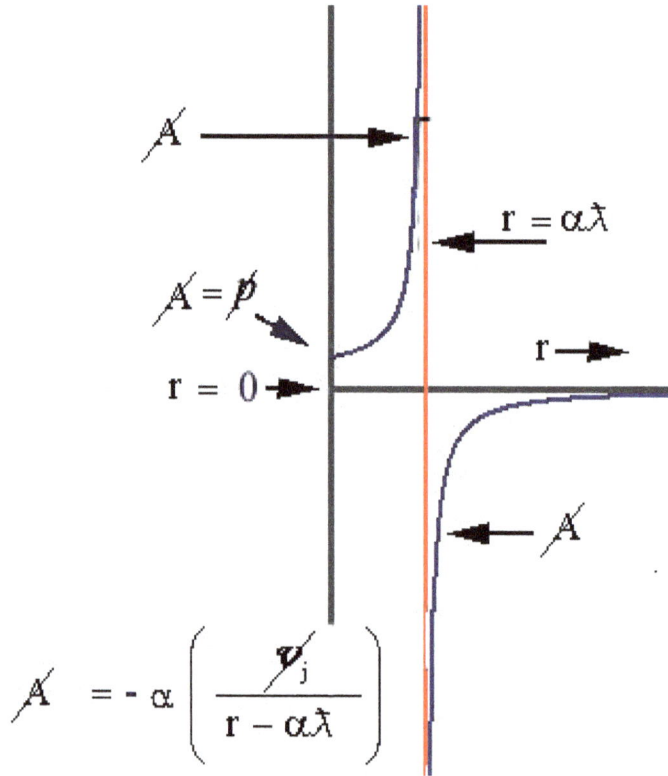

T Plot of modified charged particle Potential

Figure 1.

The αD term in Eq. (21), is the classical electron radius, and it can be noted that at the center of the particle $r = 0$, the value of A' is not infinite, but the negative of the four-momentum for that particle, or:

$$-A'_n \big|_{r_n=0} \to m v'_n = p'_n \qquad (22)$$

Thus, the sum of the gauge potentials of a collection of particles, evaluated at a single n particle:

$$\left(cm\dot{v}_n - \sum_j \not{A}_j \right)\bigg|_{r_n=0} = \left(c\not{P}_n - \sum_j \not{A}_j \right)\bigg|_{r_n=0} \qquad (23)$$

This is the canonical momentum for the n particle. The magnitude has the appearance of a Lagrangian for a particle in a collection of other charged particles. Both terms are covariant, and thus the sum is covariant. [6]

Noting that the standard classical Lagrangian for a particle in an electromagnetic field is:

$$L = -m_0c^2\sqrt{1-\frac{v^2}{c^2}} + Q(\phi - A\mathbf{g}v) = -m_0c^2\sqrt{1-\frac{v^2}{c^2}} + \not{A}\mathbf{g}\dot{v} \qquad (24)$$

This is not a covariant function but is shown to yield a relativistic action.[9]

It is proposed that instead of Eq.(24), an acceptable relativistic Lagrangian for a particle in a collection of charged particles, which is covariant, is the magnitude of Eq.(23)

$$L_n = \left| c\not{p}_n - \sum_j \not{A}_j \right| \qquad (25)$$

The vector function is just the canonical momentum, and the n particle selection can be the Lagrangian for any of the particles in the system. (See **Note 1,** for discussion of this selection)

More explicitly in velocity terms Eq.(25), is:

$$L_n = \left| \pm_n cm_n \not{v}_n - \sum_m \pm_m \alpha\left(\frac{\not{v}_j}{r_j - \alpha D_j} \right) \right| \qquad (26)$$

The velocities are geometrical fourspace vectors. (This should not be confused with an integral of a Lagrange field density.)

D. Total Lifetime Bivector Action

This section will define the total spacetime action of a particle from its initial existence in the universe to its present event.

Inserting the invariant Lagrangian of Eq. (25), into the spacetime action of Eq. (19), gives for the action:

$$\left(S_n\right) = -\left.\int_{t_i}^{t_f}\left| c\, \not{p}_n - \sum_j \not{A}_j \right| dt\, d^3x \right.\tag{27}$$

As noted in Eq. (22), the center of the n particle is the endpoint.

As was noted for Eq. (17), the action for the particle is to be summed over the interval from the initiation of the system, (Big Bang) to its current event. This interval is expanding at c and thus must be a null vector, consistent with our solution for the amplitude function, EQ. (11). Since the particle is moving at the velocity of light with respect to the initial event the time and space integrals are on coordinate time and the sum of the amplitude action is therefore, zero.

$$\not{\eta}T = \;=\left(\gamma^0 + \overset{r}{\eta}\cdot\overset{r}{\gamma}\right)T \;=\left(\gamma^0 T + \gamma^k X_k\right)\tag{28}$$

Where T is the interval time and X is the observed radius of the universe, now 13.8 billion years. Then:

$$\left(S_n\right) = \left|\left(\gamma^0 T + \gamma^k X_k\right)\left(\not{p}_n - \sum_j \not{A}_j\right) - S_0\right|\tag{29}$$

Don't need the S0

$$\left(S_n\right) = \left|\left(\gamma^0 T + \gamma^k X_k\right)\left(\not{p}_n - \sum_j \not{A}_j\right) - S_0\right|$$

$$\rightarrow \gamma^\mu X_\mu\, \not{p}_n \rightarrow \gamma^\mu X_\mu\, g\not{p}_n + I\left(\gamma^\mu X_\mu \times \not{p}_n\right)$$

$$\left(S_n\right) = \left|\gamma^\mu X_\mu\, g\not{p}_n + I\left(\gamma^\mu X_\mu \times \not{p}_n\right)\right|$$

S is the magnitude of the lifetime action of a particle from inception to the current event, and S_0 is the initial value of the action.

It has been presumed that the particle has followed a path through 4-space that minimizes the space-time interval from the initial event to the current event and that the classical vector integral over a flat Minkowski geodesic is that path.

The frame of the action is the frame in which the Big Bang is the zero point, at rest in space, and the particle is traveling at c. If the action over the life of the particle is to be calculated, the initial frame has to be the proper starting point. The particle is moving along the light cone, a null vector, in a direction away from the BB point. Relativistically the particle is moving at c respect to the BB frame, and proper time and relative motion is stopped, thus there is no contribution of the local relative velocity to the action integral. The action can thus be trivially integrated.

Initial value

Without orientation provided by the pseudoscalar of the fourspace, the action of a particle can have no spin. It is therefore asserted that the initial action at T = 0 must contain the unit fourspace orientation vector I. The pseudoscalar is represented in fourspace geometrical algebra by $I = \gamma^1 \gamma^2 \gamma^3$ which serves the role of i in complex bivector spaces. In this case it is a unit orientation vector and not the spin per se. This can be noted from its role in vector rotation.

$$e^{I\varphi} = \cos(\varphi) + I\sin(\varphi) \tag{30}$$

The initial action for a half spin particle should be the spin or 1/2ħ so S_0 will be set as 1/2 the unit bivector,

$$S_0 = \frac{1}{2} I \gamma^k \tag{0.31}$$

γ^k is the initial direction.

The total lifetime particle action, S', then is the complex bivector.

$$S_n = \left(\gamma^0 T + \gamma^k X_k\right) \left\| \left(c\not{p}_n - \sum_j \not{A}_j\right) - \frac{1}{2} I \gamma^k \right\|$$

$$S_n{}^2 = \left(\left(\gamma^k X_k\right)\left(c\not{p}_n\right) - \frac{1}{2} I \gamma^k\right)^2$$

$$= \left[\left(\gamma^\mu X_\mu\right)\left(c\not{p}_n\right)\right]\left[\left(c\not{p}_n\right)\left(\gamma^\mu X_\mu\right)\right] \rightarrow \left(\gamma^\mu X_\mu\right)\mathbf{g}\left(c\not{p}_n\right) \tag{32}$$

$$+ \left(\gamma^k X_k\right)\left(c\not{p}_n\right)\frac{1}{2} I \gamma^k + \frac{1}{2} I \gamma^k \left(\gamma^k X_k\right)\left(c\not{p}_n\right) \rightarrow I\left(\gamma^k X_k\right) \times \left(c\not{p}_n\right)$$

$$S_n \rightarrow \left(\gamma^\mu X_\mu\right)\left(c\not{p}_n\right) \rightarrow \left(\gamma^\mu X_\mu\right)\mathbf{g}\left(c\not{p}_n\right) + I\left(\gamma^\mu X_\mu\right) \times \left(c\not{p}_n\right)$$

This is the lifetime fourspace action of a charged particle existing at the current time.

In terms of the standard view of action, this would be:

$$S \rightarrow \left| S - \frac{1}{2} I \right| \tag{33}$$

For the purposes of the next section it is noted that the invariant magnitude of a velocity-position bivector AB is:

$$|AB|^2 = ABBA \tag{34}$$

For simplicity it will be designated as the square, [7] thus:

$$\left(iS_n\right)^2 = \left[\left(\gamma^0 T + \gamma^k X_k\right)\left(c\not{p}_n - \sum_j \not{A}_j\right) + \frac{1}{2} I \gamma^k\right]^2 \tag{35}$$

or heuristically can be expressed as:

$$\left(S'\right)^2 = \left(S^2 + iS - \frac{1}{4}\right) \tag{36}$$

The ¼ will vanish

The linear action will just be the cross term or the spin angular momentum

The dit is the line integral of the action

300

The square of the magnitude defined in Eq. (36), is scalar but not just the square of the action between two events, but also includes a linear component representing the phase difference between the initial and final event.

In particle velocity terms the square of the lifetime action of the n particle in the potential of all the other charged particles, Eq. (35), is:

$$\left(iS'_n\right)^2 = -\left[\left(\gamma^0 T + \gamma^k X_k\right)\left(cm_n \not{v} - \sum_m \pm \frac{\alpha \not{v}_j}{r_j}\right) + \frac{1}{2} I \gamma^k\right]^2 \quad (37)$$

I. Proposed Complete Solution, The Systemfunction

This section constructs a point solution that properly models both the phase and amplitude of a particle existing in a collection of other particles originating in an initial single event, and satisfying both the Klein-Gordon and Dirac equations.

Focusing on the action path solutions to the Schrödinger equations, Eq. (15), and noting that a particle can be considered as an energy level or state of the system. It is presumed that the initial state for a particle in the Big Bang is a constant. $(x, T = 0,0)$.

$$\psi\left(x_{1i}, t_{1i} \cdots x_{ni}, t_{ni}\right) = \text{constant} \quad (38)$$

Thus the primary constituent of the wavefunction must be the propagator that evolves the state to the final state. For the classical path Eq. (15), would be:

$$\psi\left(x_f, t_f\right) = \int Dx\left(t\right) e^{i\frac{S}{h}} \to e^{i\frac{S}{h}} \quad (39)$$

An observation of Eq. (36), shows an action square which has both the linear phase relation of the action, and the real square of

301

the action which must be related to the relativistic mass. It is proposed that the square of this is the basis of a propagator action that includes more than just the standard first order terms; there are also real terms that set the value of the amplitude factors of the solutions to the KG equation.

It is proposed that the Systemfunction point solution, $\tilde{\Theta}$, for a particle satisfying Eq. (7), and Eq. (14), is:

$$\tilde{\Theta}_n = Ae^{(S')^2 \big|_{r_n=0}} = Ae^{\left(S^2 + iS - \frac{1}{4}\right)\big|_{r_n=0}}, \qquad (40)$$

Note that the phase part of the function is scalar, linear, and somewhat conventional, but the S^2, is real, and constitutes the amplitude of the particle function. The $r_n = 0$, indicates endpoint action evaluated at the center of the n particle.

The Systemfunction then from Eq. (37), for the single n particle is:

$$\tilde{\Theta} = Ae^{\left(\left(T^2 - X^2\right) \times \left(c p_n - \sum_j A_j\right) \times \left(c p_n - \sum_k A_k\right) + I\gamma^k T \not{n} \bullet \left(c p_n - \sum_j A_j\right)\right)}$$

(41)

The function will be shown to be a solution of the KG Eq. (7), and the phase factor is a solution to Eq. (14), and the amplitude is a solution to Eq. (11).

Note that the first term, which is the square of the real action, is zero $\not{n}\not{n} = 0$, Making the real term a ratio of the space and time functions a constant equal to one.

$$\tilde{\Theta} = A \; \frac{e^{\left(T^2 \left(c p_n - \sum_j A_j\right) \times \left(c p_n - \sum_k A_k\right)\right)}}{e^{\left(X^2 \left(c p_n - \sum_j A_j\right) \times \left(c p_n - \sum_k A_k\right)\right)}} \times e^{\left(I\left(\gamma^0 T + \gamma^k X_k\right) \bullet \left(c p_n - \sum_j A_j\right)\right)}$$

(0.42)

302

Note, also that at the center of the function $T^2 = X^2$, the value of the real function is A, and at a distance Δx from the center, with the velocity equal to zero the observed function is:

$$\tilde{\Theta}_{\Delta x} = Ae^{-\left(\frac{Mc}{\hbar}\Delta x\right)^2} \tag{0.43}$$

This function describes a real particle having a spherical Gaussian shape with the proper dimensions for a classical particle such an electron.

The phase function of Eq. (41), comes from:

$$\frac{I\gamma^k}{2}\left(\gamma^0 T + \gamma^k X_k\right)\left(c\not{p}_n - \sum_j \not{A}_j\right) + \left(c\not{p}_n - \sum_k \not{A}_k\right)\left(\gamma^0 T + \gamma^k X_k\right)\frac{I\gamma^k}{2}$$

$$= I\gamma^k\left(\gamma^0 T + \gamma^k X_k\right)\bullet\left(c\not{p}_n - \sum_j \not{A}_j\right)$$

$$\tag{44}$$

The first term in the exponent of Eq. (41), is real, scalar, and related to the classical properties including particle mass ratios. The product $\not{n}\not{n}$ is the square of a null vector and always zero, and thus the exponential does not run away. The Second term is imaginary, scalar, and generates the phase or QM properties of the function.

It is strange that the KG differentials operating on the first term are not also zero, but since the differentials are a difference between the time and space coordinates, rather than a sum, it has a value.

The $\not{n}$ vector is not a field variable as would be the case for a Dirac particle wavefunction, but is the null location vector of the particle with respect to the initial event.

II. Separating The Systemfunction

The System function $\tilde{\Theta}$ for the n particle defined in Eq. (40), as shown earlier, Eq. (8), can be separated into a product of amplitude and phase functions, and the single equation can be separated into amplitude and phase equations.

303

$$\tilde{\Theta} = \Theta_R \Theta_I \qquad (45)$$

From Eq.(41), the amplitude scalar Systemfunction is:

$$\tilde{\Theta}_R = Ae^{\left(\gamma^0 T + \gamma^k X_k\right)^2 \times \left(c\not{p}_n - \sum_j \not{A}_j\right) \times \left(c\not{p}_n - \sum_k \not{A}_k\right)} \qquad (46)$$

The scalar phase Systemfunction for the n particle and solution to Eq.(14), is:

$$\tilde{\Theta}_I = Ae^{IT\not{n}\cdot\left(c\not{p}_n - \sum_j \not{A}_j\right)} \qquad (47)$$

III The First Order Complex Function Properties

This section shows the phase term is a solution to the first order equation.

Letting $I = i$ for familiarity, and $T = t$, be the local time, since the phase is cyclic, the scalar complex portion of the function of Eq.(47), is then:

$$\tilde{\Theta}_I = Ae^{it\not{n}\cdot\left(c\not{p}_n - \sum_j \not{A}_j\right)} \qquad (48)$$

Noting that:

$$t\not{n} = \left(ct\gamma^0 + \gamma^1 x + \gamma^2 y + \gamma^3 z\right) \qquad (ct)^2 = |r|^2 \qquad (49)$$

$$t\not{n}\cdot\not{p} = m\not{n}\cdot\not{v} = Et + p\cdot\vec{r} \qquad (50)$$

304

The free particle part of the exponent in Eq.(48), becomes identical to the Dirac free particle wavefunction:

$$\tilde{\Theta}_I = Ae^{i\left((Et+p\bullet\vec{r}) - \sum\left[t\cancel{n}\bullet\sum_j \cancel{A}_j\right]\right)}$$

(51)

If the potential term is ignored the form of the phase Systemfunction is identical to the Dirac wavefunction for a massive free particle. The difference is that the Systemfunction is the value of the function only at an event in spacetime, at the center of a particle, whereas the Dirac solution is the probability amplitude throughout spacetime. In the Dirac case the solution is a function of r, whereas the r in the Systemfunction is the particle location.

A. Gauge Potential Equivalence

The notable difference in The Systemfunction, Eq.(51), and the Dirac wavefunction is the presence of the potential in the function. This can be shown to be equivalent to the Dirac solution of an equation with the potential operator included.

Defining the phase function of Eq.(45), $\cancel{\partial}\tilde{\Theta}_I = \cancel{\partial}\tilde{\Theta}_a\tilde{\Theta}_b$ as, where the potential function for the j particle is a separate function:

$$\tilde{\Theta}_b = \exp\left[\mp i\sum\left(t\cancel{n}\bullet\cancel{A}_j\right)\right]$$

(52)

Noting the explicit form of the null vector $t\cancel{n}$ Eq.(49), multiplying and taking the derivative the result is:

$$\cancel{\partial}\left[\mp it\cancel{n}\bullet\cancel{A}_j\right] = \pm i\cancel{A}_j$$

(53)

(The more conventional form of $\cancel{A}$ would have Q $\cancel{A}$.)
Applying the chain rule to the scalar Systemfunction:

$$\cancel{\partial}\tilde{\Theta}_I = \cancel{\partial}\tilde{\Theta}_a\tilde{\Theta}_b = \tilde{\Theta}_b\left(\cancel{\partial}\tilde{\Theta}_a\right) + \tilde{\Theta}_a\left(\cancel{\partial}\tilde{\Theta}_b\right) = \tilde{\Theta}_b\left(\cancel{\partial}\tilde{\Theta}_a\right) + \tilde{\Theta}_a\tilde{\Theta}_b\left(iQ\cancel{A}\right)$$

(54)

305

Thus:

$$\tilde{\Theta}_b \left[-i\not{\partial} + \not{A}_j \right] \tilde{\Theta}_a = m_0 \, \tilde{\Theta}_a \tilde{\Theta}_b \qquad (55)$$

or:

$$\left[-i\not{\partial} + \not{A}_j \right] \tilde{\Theta}_a = m_0 \, \tilde{\Theta}_a \qquad (56)$$

The phase Systemfunction with the potential included in the solution has equivalence with the Dirac wavefunction, as a solution to the Dirac operator, with a potential operator.

The phase Systemfunction is thus a solution to the first order Dirac equation.

IV Observation Field

It is undisputable that the wave structure associated with the Dirac and Schrodinger first order equations are associated with probability amplitude, and it is well known that some of the associated negative wave structures are not related to probability amplitudes. The following is a presentation of a wave structure associated with the massive point particle defined here that exhibits the known features and interaction properties

The function value of Eq.(47), at the $r_n = 0$ position is can be observed in the future along the direction of the same null vector locating the particle in the initial event coordinate system. The direction to the initial event is ubiquitous, meaning that the Big Bang is at 13.8B light years distant, but the direction is arbitrary over 4 pi steradians. If an arbitrary null interval is added to the null location vector the, value of the Systemfunction event at the particle can be observed at a distant point at a future time, which is at the time of arrival of light from the event.

$$\not{\pi}T \rightarrow \not{\pi}(T + r_0) \qquad (57)$$

The value depends on the angle, distance, and time from the particle, and can be referred to as the "observation field" of the point function

Figure 4 shows the n particle with the κ event located at the center $r_n = 0$, as observed at another event. The value of the Systemfunction, which is a point function, is viewed along the null observation vector $\not{\eta} r_0$, and has a different value depending on distance and direction from the event.

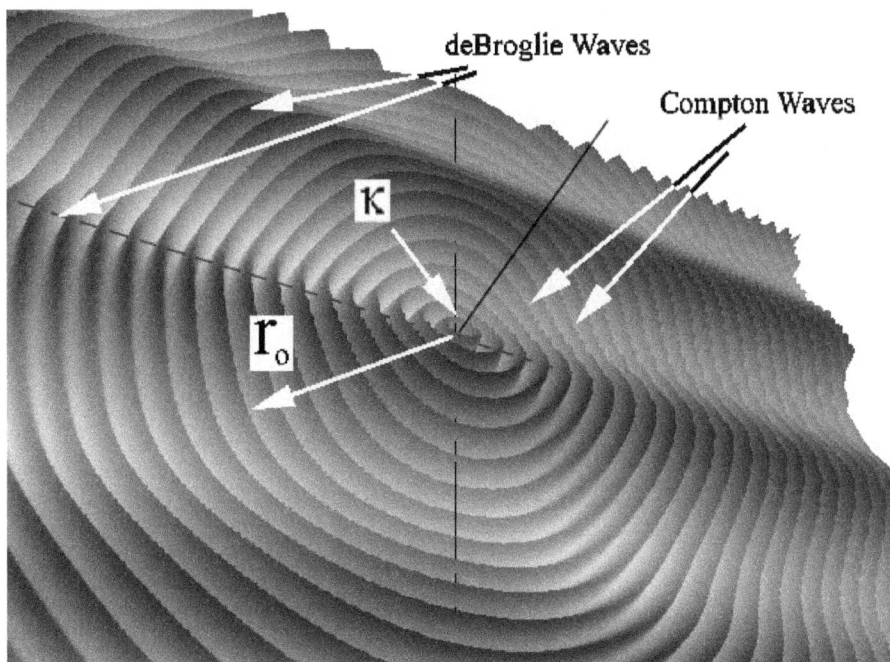

deBroglie Waves

Compton Waves

κ

r_0

Observed value of Θ_I function evaluated at n particle as viewed at r_0

Figure 4.

This is the observation of the value of the function at the κ event as transmitted by a photon. There is no physical field, or any substance defined, at that point. r_0. This observation field would represent the interaction of the point particle with another particle, as communicated by a photon from the point particle.

V The Classical Kg Equation

Putting the Systemfunction, $\tilde{\Theta}$ into the KG equation, results in the standard coordinate free equations of motion for a charged particle in the presence of the electromagnetic field of other particles, and illustrates that the same Systemfunction is not only a solution to the Dirac Equation but also the Klein Gordon equation.

Putting the complete Systemfunction of Eq.(41):

$$\tilde{\Theta} = Ae^{\left|\left(T^2-X^2\right) \times \sum_{j=1}^{N}\left(c\not{p}_n - \sum_j \not{A}_j\right) \times \sum_{k=1}^{N}\left(c\not{p}_n - \sum_k \not{A}_k\right) + i\sum\left(\gamma^0 T + \gamma^k X_k\right) \cdot \not{A}_j\right|_{r_n=0}} \tag{58}$$

into Eq(7),

$$\left(\not{\partial}^2 + m_0^2\right)\tilde{\Theta} = 0, \tag{59}$$

and executing the differential, gives:

$$-\left(c\not{p}_n - \sum_j \not{A}_j\right)\left(c\not{p}_n - \sum_k \not{A}_k\right) = -m_0^2 \tag{60}$$

Expanding Eq.(60), and using notation in Eq.(35),

$$-\left(\left(cm_n\not{\partial}\right)^2 + \left(cm_n\not{\mathcal{V}}_n\right)\sum_j \not{A}_j + \sum_k \not{A}_k \vec{S}_n\left(cm_n\not{\mathcal{V}}_n\right) \cdots\right) = -m_0^2 \tag{61}$$

Explicitly putting in the potential from Eq.(21), for the m particles this is:

$$-\left(cm_n\not{\partial}_n\right)^2 + \sum_j \pm\pm 2m_n \frac{\alpha}{r_j}\not{\partial}_j \cdot \not{\partial}_n = -m_0^2 \tag{62}$$

Or in more elementary terminology:

$$\left(\frac{m_{n0}c}{\hbar}\right)^2 = \left(\frac{m_n c}{\hbar}\right)^2\left[\left(1 - \frac{\vec{v}_n^2}{c^2}\right) - \sum_{n\ j}\pm\pm \frac{2}{m_n c^2}\frac{Q^2}{r_j}\left(1 - \frac{\vec{v}_j}{c}\cdot\frac{\vec{v}_n}{c}\right)\right] \tag{63}$$

Taking the square root gives the familiar classical linear Lagrangian expression for a particle the presence of other electrically charged particles. Note that the proper electromagnetic interaction between the particles could not have been possible without the square of the action of Eq.(35).

The square root of Eq. (63), is:

$$m_{n0}c^2 \approx + \left[m_n c^2 \left(1 - \frac{1}{2}\vec{v}^2_n \right) \mp \sum_j \frac{Q^2}{r_j} \left(1 - \vec{v}_j \bullet \vec{v}_n \right) \right] \quad (64)$$

Note this equation is the particle-particle electromagnetic interaction Lagrangian, and the classical coordinate free equation of motion for the particle. (See Appendix II) The amplitude of the function is therefore just the classical mass of a particle in an electromagnetic field. This equation is the core of the Bohr-Sommerfeld model of the atom, thus showing that the Systemfunction contains both the Quantum and Classical descriptions of particles.

The amplitude factor of the solution thus has the proper classical values of the electromagnetic interactions, but the phase factor of the Systemfunction demonstrated earlier, sets the quantization rules and values.

It is of interest to note that the next term from the expansion of Eq. (21), in the bracket of Eq. (64), is:

$$\frac{Q^2}{r} \left(\frac{\alpha \lambda_j}{r} \right) \left(\frac{\vec{v}_n \bullet \vec{v}_j}{c^2} \right) \quad (65)$$

This is the classical value of the coefficient of the Larmor spin orbit interaction energy interaction which is:

$$\Delta \boldsymbol{H}_L = \frac{Q^2}{r} \left(\alpha \frac{\lambda_e}{r} \right) \left(\frac{1}{n} \frac{\vec{v} \times \vec{r} \bullet \boldsymbol{S}}{vr\hbar} \right) \quad (66)$$

This illustrates that the classical amplitudes do not contain the QM detail available in the multiple phase solutions of the linear form.

From the foregoing, it has been shown that the real amplitude factor of the Systemfunction is a solution to the KG equation, and thus the Systemfunction is a solution to both the quadratic and linear forms of the KGD equation.

Conclusion

An alternate point particle solution has been presented that is a solution to both the KG and Dirac expressions. It is only viable in the context of the particle system as a whole, with the initial event with T = 0 (Big Bang). The solution defines both classical (amplitude), and quantum mechanical, (phase) properties of charged particles originating in a system of similar particles heretofore not connected. This is a straightforward arithmetical approach in geometric algebra to a solution of the equation in a universe that is expanding at c. The constructs and observables are outside the normal QM parameters, and thus do not conflict with standard QM methods or results, but adds another perspective. It is presumed that the defining the connection between the real amplitude and the imaginary phase factors of a complete wavefunction can be used in defining particle mass ratios.

Référence :

1. Feynman, Hibbs, 1965, Quantum Mechanics and Path Integrals McGraw-Hill pp 24,

"Today, any general law that we have been able to deduce from the principle of superposition of amplitudes, such as the characteristics of angular momentum, seem to work. But the detailed interactions still elude us. This suggests that amplitudes will exist in a future theory, but their method of calculation may be strange to us."

2. Notes on Quantum Mechanics, K. Schulten, 2000Beckman Institute University pp324 http://www.ks.uiuc.edu/Services/Class/PHYS480/qm_PDF/QM_Book.pdf

3. Janos Polonyi, 1976Path Integral for the Dirac Equation, arXiv:hep-th/9809115v1

4. G. J. Papadopoulos, J. T. Devreese, Path-integral solutions of the Dirac equation Phys. Rev. D 13, 2227–2234

5. P. A. M. Dirac, 1958The Principles of Quantum Mechanics, 4th ed., Oxford University Press, London

6. John David Jackson, 1952 Classical Electrodynamics Second Edition pp 407

7. Chris Doran, 2003 Geometric Algebra for Physicists, Cambridge University Press,

8.Ann T Nelms, Graphs of the Compton energy-angle relationship and the Klein-Nishina formula, University of Michigan Library.

9. Jon Magne Leinaas, Classical Mechanics and Electrodynamics Lecture notes,
http://www.uio.no/studier/emner/matnat/fys/FYS3120/v10/
undervisningsmateriale/ LectureNotes3120.pdf

Appendix I

Definitions, Notation, and Conventions

Feynman slash notation:

$$\slashed{A} = \gamma^\mu A_\mu \qquad \slashed{a}\slashed{a} = a^\mu a_\mu = a^2 \qquad \slashed{a}\slashed{b} + \slashed{b}\slashed{a} = 2a \cdot b$$

The radius of particle system

$$\mathfrak{R} = cT = \mathfrak{R}_0 + ct$$

Four velocity

$$\gamma^\mu v_\mu = \slashed{v}$$

Three velocity

$$\gamma^k (v)_k = \vec{v} = \gamma^k \cdot \vec{v}$$

Null unit vector

$$\slashed{\eta} = {}^{\mu}\gamma\eta_{\mu} = \left(\gamma^0 + \vec{\eta}\cdot\vec{\gamma}\right) \qquad \vec{\eta}\cdot\vec{\eta} = -1$$

Mass in this paper

$$m = \frac{mc}{h} = \frac{1}{D}$$

Rest mass

$$m_0 = \frac{m_0 c}{h}$$

Compton radius

$$D = \frac{h}{mc}$$

Vector 4 potential

$$\slashed{A}$$

Most equations

$$h = c = 1$$

Charge sign of j particle

$$\underset{j}{\pm}$$

Derivatives in slash notation

$$\gamma^{\mu}\frac{\partial}{\partial(x)_{\mu}} = \gamma^{\mu}\partial_{\mu} = \slashed{\partial}$$

Local reference coordinates x

Initial event coordinates X

The Weyl representation of the Dirac gamma matrix:

$$\gamma^1 = \begin{bmatrix} & & & +1 \\ & & +1 & \\ & -1 & & \\ -1 & & & \end{bmatrix} \quad \gamma^2 = \begin{bmatrix} & & & -i \\ & & i & \\ & i & & \\ -i & & & \end{bmatrix} \quad \gamma^3 = \begin{bmatrix} & & 1 & \\ & & & -1 \\ -1 & & & \\ & 1 & & \end{bmatrix} \quad \gamma^0 = \begin{bmatrix} & & 1 & \\ & & & 1 \\ 1 & & & \\ & 1 & & \end{bmatrix}$$

$$\gamma^1\gamma^1 = -1, \quad \gamma^2\gamma^2 = -1, \quad \gamma^3\gamma^3 = -1, \quad \gamma^0\gamma^0 = +1$$

Four velocity

$$\not{v} = \left(\gamma^1 v_x + \gamma^2 v_y + \gamma^3 v_z + \gamma^0 c\right)$$

The product of two unitless four velocities:

$$\not{v}_n \not{v}_m = \left(\gamma^1 v_{xn} + \gamma^2 v_{yn} + \gamma^3 v_{zn} + \gamma^0 c\right)\left(\gamma^1 v_{xm} + \gamma^2 v_{ym} + \gamma^3 v_{zm} + \gamma^0 c\right)$$

or

$$\not{v}_n \not{v}_m = \left[\vec{v}_n g \vec{v}_m + \bar{\sigma}\left(\vec{v}_n \times \vec{v}_m\right) + \gamma_4 c\left(\vec{v}_m - \vec{v}_n\right)\right] \quad \bar{\sigma} = \gamma^1\gamma^2\gamma^3$$

The inner product:

$$\not{v}_n \not{v}_m + \not{v}_m \not{v}_n = 2\vec{v}_n g \vec{v}_m$$

The outer product:

$$\not{v}_n \not{v}_m - \not{v}_m \not{v}_n = \left[2\bar{\sigma}\left(\vec{v}_n \times \vec{v}_m\right) + 2\gamma_4 c\left(\vec{v}_m - \vec{v}_n\right)\right]$$

Appendix II

(*Standard units*)

Particle Interaction Lagrangian

From Eq.(64), the interaction of a charged particle in an electromagnetic field of a charged particle j is:

$$m_{n0}c^2 = + \left[m_n c^2 - \frac{1}{2} m_n \vec{v}^2_n \pm \pm_{n \, J} \left(\frac{Q^2}{r_J} \left(1 - \frac{\vec{v}_J}{c} \cdot \frac{\vec{v}_n}{c} \right) \right) \right] \qquad (1.1)$$

So the interaction Lagrangian is:

$$L_{Int} = \pm \pm_{n \, J} \left(\frac{Q^2}{r_J} \left(1 - \frac{\vec{v}_J}{c} \cdot \frac{\vec{v}_n}{c} \right) \right) \qquad (1.2)$$

From Jackson p407 [6], the interaction Lagrangian for two particles is:

$$L_{int} = \frac{Q}{m_0 c} (p \cdot A) = Q \left(\not{v} \cdot \not{A} \right) \qquad (1.3)$$

With:

$$m_0 = m \sqrt{1 - \left(\frac{v}{c} \right)^2} \qquad (1.4)$$

From Jackson, [6], the Potentials for the classical electrodynamics interactions are: (*not* the quantum electrodynamics interactions).

$$A = \frac{Q\vec{v}}{r} \qquad (1.5)$$

Thus

$$L_{int} = Q\frac{\vec{v}_n}{c}\bullet A_J = Q\left[\gamma^0 + \gamma^k\left(\frac{v_n}{c}\right)_k\right]\bullet\left[\frac{Q}{r_J}\left(\gamma^0 + \gamma^k\left(\frac{v_J}{c}\right)_k\right)\right] = \frac{Q^2}{r_J}\left(1-\left(\frac{v_J}{c}\right)\bullet\left(\frac{v_n}{c}\right)\right)$$

$$(1.6)$$

and is the same as our interaction Lagrangian in the classical particle Eq.(1.2).

Note 1

Alternate Electromagnetic Lagrangian

The Standard classical non-covariant Lagrangian for a particle in an electromagnetic field is Eq.(24), and using our defined notation Eq.(21), this is:

$$L = -m_0 c^2\sqrt{1-\frac{v^2}{c^2}} + Q(\phi - A\bullet v) = -m_0 c^2\sqrt{1-\frac{v^2}{c^2}} + A\bullet\vec{v} \quad (2.1)$$

Letting $m_0 / m = 1/\gamma \rightarrow m_0 = m/\gamma$, and expanding:

$$L = -mc^2 + mv^2 + A\bullet\vec{v} \qquad (2.2)$$

In this proposal, the Lagrangian is the magnitude of the invariant gauge potential evaluated at the particle.

$$L_n = -\left|c\vec{p}_n - \sum_j A_j\right| \qquad (2.3)$$

The first order terms of this are:

$$L_n^2 = \left|c\vec{p}_n - \sum_j A_j\right|^2 = c^2\vec{p}_n^2 - 2c\vec{p}_n\bullet\left(A_j\right) + (A)^2 \qquad (2.4)$$

315

Taking square root of product;

$$L_n = -\sqrt{\left(mc^2\right)^2\left(1-\frac{v^2}{c^2}\right)\pm 2mc^2\,\cancel{v}\cdot A+\left(A\right)^2}\qquad(2.5)$$

$$L_n = -\left(mc^2\right)\sqrt{\left(1-\frac{v^2}{c^2}\right)\pm\frac{2}{mc^2}A\cdot\cancel{v_n}+\frac{\left(A\right)^2}{\left(mc^2\right)^2}}\qquad(2.6)$$

Since most of the terms are small the square root is:

$$L_n = -\left(mc^2\right)\left(1-\frac{1}{2}\frac{v^2}{c^2}\right)\mp A\cdot\cancel{v_n}+\frac{1}{2}\frac{\left(A\right)^2}{mc^2}\qquad(2.7)$$

$$L_n = -mc^2+\frac{1}{2}mv^2 \pm A\cdot\cancel{v_n}-\frac{\left(A\right)^2}{2mc^2}\qquad(2.8)$$

Comparing the classical version Eq. (2.2), with the proposed relation Eq. (2.8), we see that the first order difference in the proposed version is that the kinetic energy enters with the classical value, whereas in the classical Lagrangian the enters with twice the classical value of the kinetic energy.

Since multiplying the Lagrangian by a constant or adding a constant, has no effect on the equations of motion, the difference between the classical and the proposed Lagrangian should have no effect on the predicted equations of motion.

It is straight forward to show the Lagrange equations of motion resulting from the proposed Lagrangian produce acceptable equations of motion for a free particle.

Momentum:

$$\frac{\partial L}{\partial \dot{x}^{\mu}}=2p\qquad(2.9)$$

Acceleration:

$$\frac{d}{dt}\left(\frac{\partial L}{\partial \dot{x}^{\mu}}\right)=0\qquad(2.10)$$

Coordinate dependence:

$$\frac{\partial \mathbf{L}}{\partial x^{\mu}} = 0 \qquad (2.11)$$

Note 2

Constraint on Amplitude Function

Starting with the factored expression:

$$\left(i\not{\partial} + m_0\right)\left(i\not{\partial} - m_0\right)\tilde{\Theta} = 0 \qquad (3.1)$$

Or:

$$\left(i\not{\partial} - m_0\right)\tilde{\Theta}_R\tilde{\Theta}_I = 0 \qquad (3.2)$$

Applying the chain rule:

$$\left(i\not{\partial} - m_0\right)\tilde{\Theta}_R\tilde{\Theta}_I = \tilde{\Theta}_R i\not{\partial}\tilde{\Theta}_I + \tilde{\Theta}_I i\not{\partial}\tilde{\Theta}_R - m_0\tilde{\Theta}_R\tilde{\Theta}_I = 0 \qquad (3.3)$$

Dividing by $\tilde{\Theta}_R$ this is:

$$\left(\frac{\tilde{\Theta}_I}{\tilde{\Theta}_R}i\not{\partial}\tilde{\Theta}_R\right) + \left(i\not{\partial} - m_0\right)\tilde{\Theta}_I = 0 \qquad (3.4)$$

Thus:

$$\left(i\not{\partial} - m_0\right)\tilde{\Theta}_I = 0 \qquad (3.5)$$

Only if:

$$\not{\partial}\tilde{\Theta}_R = 0 \qquad (3.6)$$

317

The Concept of Mass as Interfering Photons

D.T. Froedge

V032615

Abstract

For most purposes in physics the concept of mass particles and photons are treated as though they are completely separate and distinct entities having little connection accept through collision interactions. This paper explores the concept of a mass particle being viewed as a pair of trapped photons in a mass-less box demonstrating proper relativistic dynamics and, Lorentz covariance. The mechanism of trapping of the photons in a particle is not herein defined and not important to the discussion since it is not required by the mechanics or mathematics that they be connected. Although this presentation is more relatable to a simple particle such as an electron, the dynamics must be the same for all mass particles with primary constituents that have phase velocities equal to c. This illustrates the concept of the equivalence of mass and energy, and why mass velocity cannot exceed the speed of light..

Introduction

This paper was originally a demonstration of the connection, or comparison of the dynamics between a pair of trapped photons in a massless box, and a massive particle. With the realization that a gradient in c can induce not only the well known photon trajectories but also the proper dynamical properties of a mass defined as trapped photons, it has taken on, a greater degree of interest.[8]. If gravitation is merely a gradient in c that can be induced by QFT effects of path integrals, the mystery may be a little less obscure.

The special theory of relative, through the Lorentz transformations, yields the energy velocity relation for photons and particles, one through a shift in frequency, the other through a shift in mass. Considering these particles as different forms of energy, bestows a distinction between the forms of energy that is possibly unwarranted. The Lorentz transforms applied to a pair of photons can be shown to yield the same results as the transforms applied to a mass particle.

Physical Model Appearance

The mechanism of containment of the light speed constituents need not be addressed for the purpose of this paper, nothing really requires it. The mathematical mechanisms are the same whether the photons are confined or not, and posing a massless box or stationary mirrors that contains the particles is as good as any other physical mechanism for the purposes here. If a photon is absorbed by an atom, there is an increase in mass equivalent to the energy of the photon. If an apparatus has a pair of parallel mirrors that trap photons the overall apparatus has an increase in mass, when they trap photons, thus there should be no issue as to whether two confined opposite going photons constitute mass.

A presumed appearance for particle such as an electron modeled by a pair of trapped opposite photons is not exactly a pair bouncing back and forth. For the simple electron, in the rest frame it is more likely the particle is spherical, with a path defined by the Dirac matrix $\gamma_1 \gamma_2 \gamma_3$ along each axis.

The angular momentum any axis being $\frac{1}{2}\hbar$. If the particle is set in motion, and presuming a standing wave in the rest frame, the part of the cycle in the direction of motion will see a decreased in wavelength, and the phase against the velocity would see an increased wavelength.

Since the direction of an observer's velocity is arbitrary, the defining of what constitutes separation into the forward and backward photons from a continuous energy loop has to do with the fact that for each observer there are two states forward and backward, along the velocity vector that share the total energy of the particle. The proportion of the energy in each state determined by the Lorentz transforms. A photon would have only one such state.

Obviously complete photons are not the constituents of particles otherwise the particle would disintegrate. Presumably, however the constituents are light speed, such as leptons, quarks, gluons, etc,

I Momentum

Consider a thought experiment, in which two photons are placed in a perfectly reflecting mass-less container. If the two photons are not aligned in the given frame, then there is some sub-light speed frame, in which the photons are aligned, and in opposite directions, as well as having equal energy and frequency. This frame is the rest frame for the center of mass of the container for the two photons.

Using the momentum for the photons to be:

$$\vec{P} = m\vec{c} = \left[\frac{h\nu}{c}\right]\frac{\vec{c}}{c} \tag{1}$$

where we can designate an energy equivalent "mass " for the photon to be:

$$m = h\nu / c^2 \tag{2}$$

The momentum of the container with respect to a moving frame of reference with velocity v is then:

$$\vec{P} = \left(m_1 + m_2\right)\vec{v} \tag{3}$$

From the perspective of the individual opposite-going photons the momentum is:

$$P = P_1 + P_2 = \frac{h\nu_1 - h\nu_2}{c} = \frac{h\Delta\nu}{c} = \frac{h}{\lambda_B} \tag{4}$$

The wavelength of the difference in the frequency here, or the "beat" frequency, is just the simple deBroglie wavelength.

The total energy, which is the sum of the energy of the photons, and thus sum of the frequencies, yields the simple Compton wavelength:

$$\frac{E_1 + E_2}{c} = \frac{h\nu_1 + h\nu_2}{c} = \frac{h}{\lambda_C} \tag{5}$$

Using the above noted designation for "mass" we can write for the total "mass":

$$m_T = (h\nu_1 + h\nu_2)/c^2 \tag{6}$$

Defining a mass for photons is not a unique concept and has been used by others [4]

The momentum is then:

$$P = m_T v = (m_1 - m_2)c \tag{7}$$

Solving for velocity:

$$\frac{v}{c} = \frac{(m_1 - m_2)}{(m_1 + m_2)} \tag{8}$$

This is notably just the velocity for the center of mass for two opposite going photons.

Since for a particle:

$$m_0^2 = m^2\left[1 - \left(\frac{v}{c}\right)^2\right] \tag{9}$$

Putting in m_T, and v/c and solving gives:

$$m_0^2 = (m_1 + m_2)^2 - (m_1 - m_2)^2 = 4m_1 m_2 \tag{10}$$

So the square of the rest mass of the particle is four times the product of the "mass" of the individual photons.

II Doppler

The same result is found by the use of the relativistic Doppler shifts on the individual photons as the result of a change in the velocity same picture can be viewed from the standpoint of the Doppler shift, on the transformation of velocity coordinates for the two photons.

The relativistic Doppler shift of the photons from one velocity frame to another is:

$$v_1{}' = v_1 \left[\frac{1 - \dfrac{v}{c}}{1 + \dfrac{v}{c}} \right]^{1/2} \qquad v_2{}' = v_2 \left[\frac{1 + \dfrac{v}{c}}{1 - \dfrac{v}{c}} \right]^{1/2} \qquad (11)$$

or using the above noted conventions for energy equivalent mass:

$$m_1{}' = m_1 \left[\frac{1 - \dfrac{v}{c}}{1 + \dfrac{v}{c}} \right]^{1/2} \qquad m_2{}' = m_2 \left[\frac{1 + \dfrac{v}{c}}{1 - \dfrac{v}{c}} \right]^{1/2} \qquad (12)$$

Multiplying the two relations gives:

$$m_1{}' m_2{}' = m_1 m_2 = \mathrm{cons\,tan\,t} \qquad (13)$$

and simple math gets:

$$m_1 m_2 = \frac{\left(m_1 + m_2 \right)^2 - \left(m_1 - m_2 \right)^2}{4} \qquad (14)$$

and:

$$\left[1 - \left(\frac{v}{c} \right)^2 \right] = \frac{4 m_1 m_2}{\left(m_1 + m_2 \right)^2} = \frac{m_0^2}{m^2} \qquad (15)$$

which is the same as the above relation, found for conformance to relativistic kinematics, the model thus transforms properly.

III Four Momentum

Another approach to illustrate the mechanics of such a particle can be done by defining the null four momentum of a pair of opposite going photons. Defining the photon mass as in Eq.(6), the null four-momentum of two photons with opposite going phase velocities in the geometric algebra matrix form is:

$$\vec{P_1} = m_1 \left(\gamma^k c_k + \gamma^0 c \right) \tag{16}$$

$$\vec{P_2} = m_2 \left(-\gamma^k c_k + \gamma^0 c \right) \tag{17}$$

The square of the sum of the two null vectors is:

$$\left(m_1 + m_2 \right)^2 - \left(m_1 - m_2 \right)^2 = 4 m_1 m_2 = m_0^{\,2} \tag{18}$$

The magnitude of each of these null four-momentum is zero for covariance, and the sum of two such moments must be constant. Thus m_0 must be invariant fixed quantity associated with the pair of opposite going photons. If this is defined as a rest mass then it is easy to identify:

$$\left(m_1 + m_2 \right)^2 = m_T \tag{19}$$

as the total mass. Factoring the total mass from Eq.(18), gives:

$$\left(m_1 + m_2 \right)^2 \left[1 - \frac{\left(m_1 - m_2 \right)^2}{\left(m_1 + m_2 \right)^2} \right] = m_0^{\,2} \tag{20}$$

Noting that from Eq.(8),:

$$\frac{\left(m_1 - m_2 \right)}{\left(m_1 + m_2 \right)} \tag{21}$$

is the ratio of the velocity of each photon to the velocity of the center of mass.

$$(m_1 + m_2) v_c = (m_1 - m_2) c \qquad (22)$$

This makes Eq. (20), the relativistic energy equation for a mass particle.

$$m^2 \left[1 - \frac{v^2}{c^2} \right] = m_0{}^2 \qquad (23)$$

It can thus be asserted that two light speed photons, or other confined light speed, zero rest mass particles, have the property of a mass particle, with mass energy equivalent to the energy of the individual particles.

V Gravitation

The paper [8] "Gravitation is a Gradient in the Velocity of Light" by the author, details the interaction of gravity with the two photon model showing that the Lorentz transform with a gradient in c accelerates the two photon model exactly as does gravitation. The details are developed in [8] and not restated here.

Conclusion

The concurring points of similarity of the opposite going photons in a massless box model and the particles are consistent with:
1) The deBroglie wavelength.
The Compton wavelength
The zero velocity rest mass
The total energy
Velocity transforms
Gravitation

All of the real internal constraints such as spin, energy, etc, which may be important to the actual mechanics of holding a particle together are not necessary to understand the concept.

Using a reflecting container is somewhat artificial, but the dynamics of the center of mass of two photons is the same whether the photons are confined or not, and the transformation of momentum between velocity frames, are not depend on the internal structure. The transformations and the mechanics for both the photon pair, and a massive particle are exactly the same, and it is easy to understand from this model why mass particles cannot exceed the speed of light.

http://www.arxdtf.org/

References:

1. DT Froedge, Scalar Gravitational Theory with Variable Rest Mass, V020914, http://www.arxdtf.org/css/grav2a.pdf

2. DT Froedge, The Velocity of Light in a Locally Conserved Gravitational Field, V101914, http://www.arxdtf.org/css/velocity.pdf

3. Roger Blandford, Kip S. Thorne, in Applications of Classical Physics, (in preparation, 2004), Chapter 26 http://www.pma.caltech.edu/Courses/ph136/yr2002/chap26/0226.1.pdf

4. Alex KruchkovBose-Einstein condensation of light in a cavity http://arxiv.org/pdf/1401.0520v2.pdf

5. D. Kharzeeva, K. Tuchinb, Vacuum Self–Focussing of Fery Intense Laser Beams, arXiv:hep-ph/0611133v2

6. Feynman, Hibbs, 1965, Quantum Mechanics and Path Integrals McGraw-Hill pp 24,

7. Chambers, R.G. (1960). "Shift of an Electron Interference Pattern by Enclosed Magnetic Flux". *Physical Review Letters* 5: 3–5.Bibcode:1960PhRvL...5....3C (http://adsabs.harvard.edu/abs/1960PhRvL...5....3C)

8. DT Froedge,. Gravity is a Gradient in the velocity of light V031515, http://www.arxdtf.org/css/Gravitation.pdf

The Flaw in Quantum Mechanics

D.T. Froedge

V012520

Abstract

There is a flaw in The Born structure of Quantum Mechanics that has continuously caused conceptual problems. This error is the central problem of the double slit, the EPR Paradox and Bell's Inequality, and leads to what Einstein called the "spooky action at a distance." The phenomenon generally interpreted as quantum entanglement, is nonlocal and violates the Lorentz faster than light communications. There is an explanation that does not involve hidden variables, superluminal communications, or spooky action at a distance that is explored in this paper.

Introduction

When Born [1][2][3], was working on the statistical interpretation of y. He at first thought that y was a probability density for particles but then he noted on a cue from Einstein that the square of the optical wave amplitudes photons was a photon density. He subsequently decided that y is amplitude of probability (whatever that meant) and set the probability density of a particle or energy density to be the value of $y * y$.

For bound States $y*y$, it is asserted here that the square is not proper, but is the product of the probability density of two independent conjugate wavefunctions, thus it is the coincidence of the probability densities. It is postulated that this should properly be $y_1 y_2$ where, y_1 and y_2 are independent conjugate particles.

If y_1 and y_2 are probability density of two density wavefunctions, then the probable coincidence location density of the of these two particles is just the product of the probabilities:

$$P = P_1 P_2 = y_1 y_2 \tag{1}$$

The integral over space of this product is a transient having no fixed value *unless y_1*, is the conjugate of $y_1* = y_2$. That is: opposite direction, and opposite phase.

In this case there is a rest frame and the presence of Lorentz invariant, energy density.

$$\int_0^\infty y_2 * y_2 \, d\tau = 1 \tag{2}$$

If the wavefunctions are conjugates then in that space there is a Lorentz invariant quantity of energy, equivalent to a rest mass that is invariant under a velocity transformation. This can be illustrated by simply noting the product of the relativistic Doppler shifted product of two opposite going photons is an invariant whereas a single photon is not.

For an atomic system this is the bound steady state condition for quantized energy states of the Schrodinger equation. The Schrodinger equation is time symmetric and thus if y is a solution then the conjugate $y*$, is also a solution, and there exists for a bound state standing waves associated with that state [4]. As is well known the energy in a bound state constitutes a rest mass addition to the rest mass of a system and thus an occupied energy state is an invariant contribution to the mass

Photon

A distinction should be made between the bound state conjugate product of independent wavefunctions, and the conjugate square of the photon. The product of two conjugate wavefunctions is a Lo-

330

rentz invariant whereas multiplying a wavefunction by its conjugate wavefunction does not make the wavefunction invariant.

For a free photon y is a probability density but there is no conjugate particle present. The product of y^*y is an energy or particle density, being analogous to squaring the velocity of a collection of particles and the wavefunction is not Lorentz invariant

Photon Atom Absorption

For a single photon, the probability density y is not a Lorentz invariant thus the space integral of y^*y is a mathematical value of the kinetic energy density but does not represent the conversion to an invariant energy.

It is postulated that for an atomic system to absorb a photon, creating an allowed Schrodinger energy state there must be the coincidence presence of a photon and it's conjugate. The single photon wavefunction cannot split and cannot exist as a solution to the Schrodinger equation.

The single photon does not have a conjugate particle present (y^*y) and cannot be an energy solution to the Schrodinger equation thus it cannot be absorbed or re-radiated by spontaneous radiation in an atomic system. The photon can kinetically interact with an atomic system by the process of Compton scattering, transferring the energy to kinetic thermal.

Photon Double Slit

A screen on which there is a projection of double slits from a coherent source there is a flux of photons going through both slits. As a single photon hits a target and reflects (Fig.1), it becomes a conjugate of the following incident particle. At that point there exists the probability of transferring the photon energy $\hbar\omega$, to an energy level of the reflecting particle inducing a full $\hbar\omega$ detection of the event. If it is a photon detector there would be an event or if it is an atomic system it could spontaneously reradiate

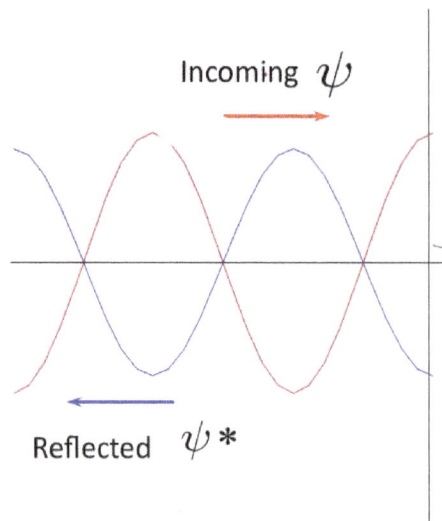

Incoming ψ

Reflected $\psi*$

Photon wave function reflecting

Fig. 1,

The probability of an energy exchange event will be proportional the phase of the conjugate photon wavefunctions. The pattern on the screen from the photons coming from two slits is then the familiar double slit pattern. The Born or Copenhagen view of this is the that the particle doesn't exist until it hits the screen and the single probability density wavefunction collapses, In this proposal there is a simple flux of photon probability densities arriving with differing phases that generate the interference pattern reflection and detection on the screen.

The mechanism of photon energy transfer in this proposal is that; two photons arriving simultaneously at a point producing an energy probability density proportional to their phase that can deliver energy to the target. Each travels separate paths, with a probability of arrival defined by Feynman path integrals, [5] and the Heisenberg uncertainty relations [6].

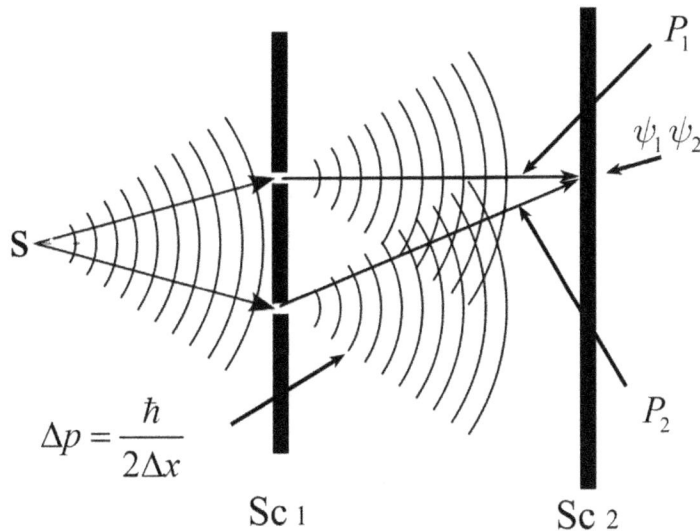

$$\Delta p = \frac{\hbar}{2\Delta x}$$

Sc 1   Sc 2

Interference of two photon fluxes of target screen

Fig. 2,

Particle Counting, Single Particle Flux Interference

In this interpretation, the number of events detected is the number of two particle coincidence events. The photon flux necessary to trigger detection is the flux necessary produce coincidences. Since the only measurable is the detection rate, the actual number of photons passing through each of the slits cannot be determined.

Conservation of energy

The transfer of energy from two coincident photons is most efficient when the detected photons are exact conjugates as in a laser beam and thus can deliver energy of $E = \hbar\omega$. Most photon encounters are not of conjugate photons, but are of only the Compton scattering which converts the photon energy to heat.

Although the out of phase photons are not visible in the double slit pattern the number of photons in the flux per steradian is the same at the minimum as the maximum. If the flux of a double slit radiation pattern is measured with a

333

blackbody thermopile pyrometer accounting for all the energy, the energy at the peaks and nulls should have the same energy flux.

Double Slit Experiment

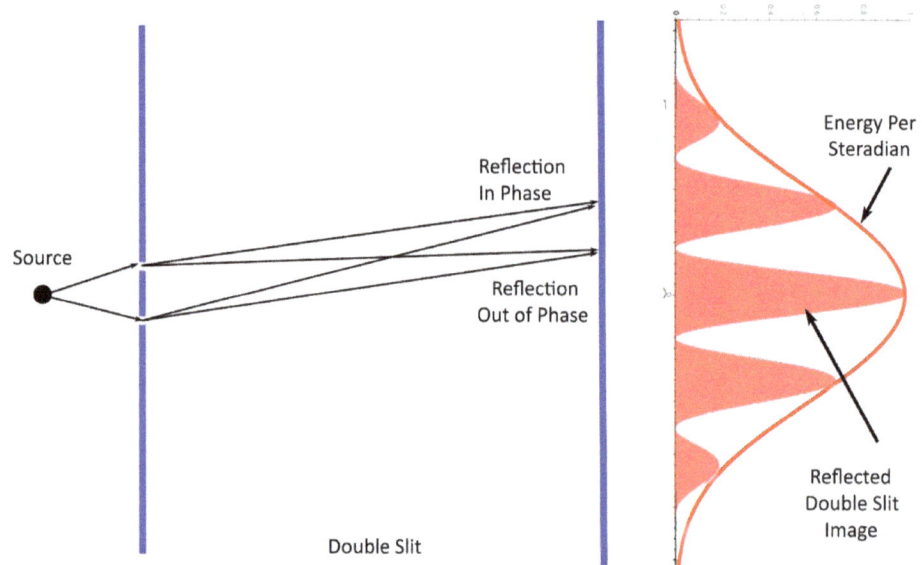

If the conjecture is correct, the energy per steradian at the target is primarily determined by the single slit as the result of the Heisenberg uncertainty relation, and should be independent of the double slit interference pattern.

Spooky Effect Explanation

Bells Inequality

The most notable experiment of QM, confirming the nonlocal aspects of QM is the EPR conjecture [7]. Measurements over the past 50 years associated with Bell's inequality have virtually confirmed that Einstein's "spooky action at a distance" is real, and somehow instantaneous communication over vast distances between measurements happens [8]. This is an illogical aspect of QM, and this alternate interpretation offers a simple explanation.

For those familiar with the measurements of Bell's Inequality test by polarization of entangle photons, the probable coincidence match between two polarizers that are set an angles $\theta_1 = 0 + \theta_0$, and $\theta_2 = 0 - \theta_0$ should logically be, for a 60^0 angle:

$$P_m = \cos(\theta_1 - \theta_2), \quad @ \; 60^0 \;\; = \;\; .5 \tag{3}$$

Whereas the actual measurement and predictions of QM and is;

$$P_{QM} = \cos^2(\theta_1 - \theta_2), \quad @ \; 60^0 \;\; = \;\; .25 \tag{4}$$

Our postulate is that a photon detection has to be the result of two photons arriving at the detection point at the same time.

If detection is only possible when two photons from the same source are coincident at the detector, then the probability of coincident is the product of the probabilities:

$$P_m = \cos(\theta_1 - \theta_2) \times \cos(\theta_1 - \theta_2) \quad\quad @ \; 60 \; = \; .25 \tag{5}$$

This yields the same results as QM without the necessity of postulating superluminal communication at a distance.

The measurements are in agreement standard QM, and with Bells expected values and thus there is no discrepancy

Conclusion

An alternate interpretation of the nature of photons and quantum mechanics has been presented that is physically mimics standard QM but explains some of the more puzzling aspects without non-local effects, hidden variables [9], or require superluminal communications. It does in some way contradicts Dirac's statement that "each photon then interferes only with itself," [10] with the double slit phenomena being an interference of two photons.

A simple experimental test using a blackbody thermopile pyrometer that can measure the energy flux across the image of a double slit pattern should be able to distinguish the physical results of this proposal, from the results expected of standard QM. This experiment is discussed in "Testing the Flaw in Quantum Mechanics"[11]

References:

1. M. Born On the quantum mechanics of collisions, (1926), Princeton University Press (published 1983). pp. 863–867. Bibcode:1926 Z Phys...37..863B. DOI: 10.1007/BF01397477. ISBN 978-0-691-08316-2

2. M. Born, L. Infeld, Die Streuung, "Von Licht An Licht", PhD thesis, 1936, Ukrainian Physico-Technical Institute; Phys. Zeit. Sow. 11, 263, 1937

3. M. Born, Statistical Interpretation of Quantum Mechanics, Science.122 issue 3172, 675-679 (1955) doi10.1126/science.122.3172.675

4. E. Schrodinger, Transl. The present situation in quantum mechanics, 1935, Translator: John D. Trimmer, Proc. of the Ame. Philosophical Society, 124, 323-38

5. Feynman, Hibbs, 1965, Quantum Mechanics and Path Integrals McGraw-Hill

6. W. Heisenberg, Physics and Philosophy: The Revolution in Modern Science. (Allen & Unwin, London, 1958

7. A. Einstein, B. Podolsky and N. Rosen, Can Quantum-Mechanical Description of Physical Reality Be Considered Complete? , 1935, Physical Review, 47 777–780.

8. J Bell. On the Einstein Podolsky Rosen paradox, Physics, 1 Nov. 1964, Physique Fizika 1, 195

9. D. Bohm, A Suggested Interpretation of the Quantum Theory in Terms of "Hidden Variables, Phys. Rev. 85 (1952) 180 - 193.

10. P. Dirac, On the theory of quantum mechanics,1926, Proceedings of the Royal Society, A 112, 661–677.

11. DT. Froedge, Testing the Flaw in Quantum Mechanics http://doi.org/10.13140/RG.2.2.24118.45128

Relativistic Time Dilation Illustrated in Delta-c Mechanics, The Twin Paradox is an Illusion

D.T. Froedge

V122423

Abstract

The twin paradox of special relativity is the curious circumstance; when two twin observers move far apart at high velocity and then come back together, one is younger than the other. An observer in one frame always sees the other's clock as having run slower and his twin as younger.

The paradoxical difference is often ascribed to the asymmetry of the acceleration, or other observational issues. If the traveler is a mass particle, which can have its velocity reversed in an instant the acceleration is not relevant to the issue.

The paradox is shown here and illustrated by aspects of Δc mechanics to be an illusion. The Illusion results from not taking into account the mass relativistic effect on the traveler in the moving frame which offsets the time difference.

Electrons as illustrated in Δc mechanics are a pair of self-bound photons rotating around the center of momentum [1], [2]. By observing the photons in the stationary particle, and in the moving particle, both in the stationary fame, there is no need to apply the Lorentz transform, and the result is quite different.

All electrons have the same mass and Compton frequency which is twice the Compton frequency of the individual rotating photons.

To evaluate the time differential in moving and stationary frames, consideration is given to two electrons being the primary clocks. By observing the mechanics of the photons in the two electrons in the same Lorentz reference frame, the time issue can be resolved.

Preliminaries

Consider a pair of electrons as timing devices or clocks, with a Compton frequency of $v_e = m_e c^2 / \hbar$, and radius, $\lambda_e = \hbar / mc_0$, in relative motion.

Since by Lorentz transform considerations, photons in all frames including the electron travel at c_0. The motion of those photons can be evaluated in the stationary frame for both the stationary and moving particles.

~

The electron has spin, so for simplicity the rotation axis can be aligned along its direction of travel. This configuration is useful in demonstrating the difference in the rate of the passage of time. It is presumed that the electrons move along their rotational axis at a velocity v.

A sketch of the trajectory motion of photons in the stationary particle (blue), and the trajectory of one of the photons in the moving electrum (red), is presented in Fig 1.

The blue circle is the trajectory of the photons of the stationary electron, and the red spiral is the trajectory of the photon in the moving electron.

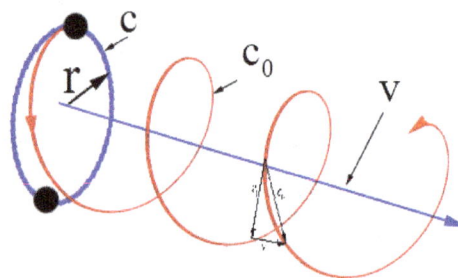

Fig. 1.

340

Compton Radii

If λ_e is the Compton radius for the electron at rest and λ'_e is the radius for the moving electron, then the respective radii are

$$\lambda_e = \frac{\hbar}{m_e c_0} \qquad\qquad \lambda'_e = \frac{\hbar}{m'_e c_0} \qquad\qquad (1)$$

The relativistic mass relation between the rest and the moving mass is:

$$m_e = m'_e \left(1 - \frac{v^2}{c_0^2}\right)^{1/2} \qquad \rightarrow \qquad m'_e = \frac{m_e}{\left(1 - \frac{v^2}{c_0^2}\right)^{1/2}} \qquad (2)$$

Thus the radius of the electron in the moving frame Eq. (2), is reduced by the relativistic Gamma factor. The reduction of the collisional radius for moving particles is a well-known experimental phenomena.

$$\lambda_e = \frac{\hbar}{m_e c_0} = c_0 t_0 \qquad\qquad \lambda'_e = \frac{\hbar}{m_e c_0}\left(1 - \frac{v^2}{c_0^2}\right)^{1/2} = c_0 t_0 \left(1 - \frac{v^2}{c_0^2}\right)^{1/2}$$

$$(3)$$

The Lorentz transforms require photons to travel at c_0 relative to an observer, thus since both electrons are observed in rest frame the velocity of both is c_0, but for the moving electron with an increased mass and reduced radius for photons of both electrons move at c_0 and complete one revolution at the same time with the same period.

For both particles the distance travelled is the same, but because of the reduced radius of the moving particle the photon is able to complete the revolution in the same time, t_0

The stationary and moving electrons can be displayed on a flatting cylinders the as shown in Fig, 1.

Graphically this is:

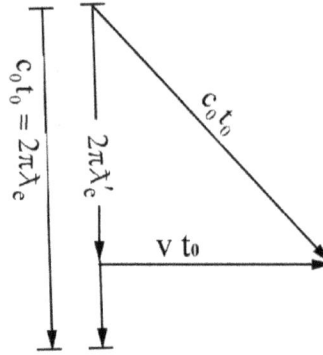

Both particles start at the top of the triangle. The photon in the stationary particle moves at c_0 and travels a distance $c_0 t_0$ in one period. The photon in the moving particles travels at c_0 with the center of mass moving perpendicular at a velocity v

For the electron at rest, the distance traveled in one period, t_0 is:

$$s = c_0 t_0 = 2\pi \lambda_e \qquad (4)$$

For the moving particle the photon velocity, c_0 is the same as the stationary particle, but it is moving horizontally with the center of mass at a velocity of v.

By the Pythagorean Theorem the distance in terms of the velocity and radius is:

$$s' = \sqrt{\left(2\pi \lambda_e'\right)^2 + \left(vt_0\right)^2} \qquad (5)$$

From Eq.(2), the distance traveled by the photon in the moving electron is:

$$2\pi \lambda_e' = 2\pi \lambda_e \left(1 - \frac{v^2}{c_0^2}\right) = c_0 t_0 \left(1 - \frac{v^2}{c_0^2}\right) \qquad (6)$$

Putting this into Eq.(5), shows the distance traveled along the trajectory of the moving photon completes the revolution around

342

the λ' radius in the same time as the electron completes the revolution around the radius λ_e, of the stationary particle.

$$s' = \sqrt{t_0^2 \left(c_0^2 - v^2\right) + \left(vt_0\right)^2} = c_0 t_0 \tag{7}$$

Conclusion

If the rotations have the same period then the photon trajectory's remain in phase, thus the number of cycles is the same. **There is no accumulated time difference.**

The time duration in the moving electron measured by the number of cycles is the same as in the time duration the rest frame, and the twin re-arriving from a distance star is the same age as the one that stayed on earth. The evaluation of this from either electron in either direction is the same, **and there is no paradoxical difference.**

The historical twin paradox has ignored the relativistic mass change to the internal clock of the traveler.

References:

1. DT Froedge, The Dirac Equation and the Two Photon Model of the Electron revised, April 2021, DOI:10.13140/RG.2.2.19095.70564, https://www.researchgate.net/publication/350922403

2. DT Froedge, The Electron as a Composition of Two Vacuum Polarization Confined Photons, April 2021, DOI: 10.13140/RG.2.2.18971.18722 , https://www.researchgate.net/publication/350740864

Early Discussions

To: DT Froedge
Subject: FW: Gravitation

From: Rai Weiss [mailto:weiss@   u]
Sent: Tuesday, July 21, 2015 6:14 PM
To: DT Froedge
Subject: Re: Gravitation

Don Travis,
It was a pleasure to meet you and, especially, to learn about your various lives. I encourge you to continue being a curmudgeon about fundamental ideas. It will keep you young.

I thought a little more about where I got stuck with your idea of the composition of a particle as a pair of photons. It isn't with the photons but I have a problem with the box. Something has got to hold the box together against the enormous electromagnetic stresses on the box. This is true whether you think of a standing waves or of two photons going in opposing directions in the box. It is the same problem that Lorentz had with the classical model of the electron, what held the charge distribution together. This was never answered until it was buried in the quantum mechanics and then later "explained" by the strong forces.
Wheeler got around the problem by having enough photons in a small region so that their self gravitation held them in a clump. This is Wheeler's geon.

The radiation reaction force. I did not draw a sufficiently convincing picture of how the retardation in the circular dipole oscillator is responsible for the central force having a small tangential component opposing the motion of the radiating charge. The radiation reaction force comes about because of the change in the direction of the acceleration of the radiator. The electric field lines originate not from the center of the system but rather from the inferred present position of the attracting charge as if it had continued to move at the velocity a time $t=r/c$ earlier. In other words, the attracting electric field is offset from the center which gives a tangential component to the electric field attracting the radiating particle. Draw a picture and it will become reasonable. So as we concluded at the end of our conversation, your version of gravity will probably also radiate gravitational waves.

RW

   On Tue, 21 Jul 2015, DT Froedge wrote:

> Dr. Weiss
>
> It was delightful to meet with you yesterday and discuss a little
> physics. I can't think of when I have enjoyed a conversation more. It
> was very gracious of you to take the time.
>
> The parking lot attendant took the money, and let me out without a hitch.
>
> Thanks again,
>

From:	Mackenzie Richard <richard.mackenzie@umontreal.ca>
Sent:	Wednesday, November 06, 2019 9:51 AM
To:	dtfroedge@glasgow-ky.com
Subject:	The Flaw in Quantum Mechanics.pdf
Attachments:	The Flaw in Quantum Mechanics.pdf; ATT00001.txt

Hi DT,

Here is your paper with a few comments written on it. I must admit I am not clear from the outset what you are getting at (sorry to say it, but your terminology is pretty fuzzy), and I didn't get very far.

As far as I can tell, your double slit explanation relies on some sort of combination of probabilities for two photons. This can't be, because the experiment shows interference if photons are sent one at a time (as I am sure you are aware).

It might be interesting to see if you can relate the probabilities of observing a photon at a point on the screen (to be precise, at a tiny area centered at the point, or else one should say the probability per unit length along the screen of observing the photon) with only one slit open, then with the other slit open, then both. You had better be able to reproduce the normal quantum mechanical result (and even then, you would face an uphill battle convincing people to adopt your approach when the standard quantum mechanical approach works just fine).

Take care,

Richard

From:	Mackenzie Richard <richard.mackenzie@umontreal.ca>
Sent:	Sunday, January 04, 2015 11:38 AM
To:	DT Froedge
Subject:	Re: path integrals

Hi again DT,

Regarding your question below: The path integral for photons is complicated due to gauge invariance, as you no doubt know. But it can be done with gauge fixing, and the net result is the photon propagator, which depends on the gauge condition; the more common ones are Lorentz and Feynman gauge. This is discussed in many text books, eg Peskin and Schroeder or Srednicki. (Both are excellent but Srednicki uses an unusual metric which is a turn-off for me!) Another very interesting discussion of related issues (but very low-tech -- it is not really path integrals per se but a discussion of amplitudes in quantum mechanics -- the underlying notion behind path integrals) is Feynman's Lectures on Physics, vol 3.

Once you know the propagator for whatever type of particle you are studying (say photons or ordinary non-relativistic massive particles), you can now address the question you describe below. I have to say I am not sure the question makes sense, or at least you have to be very careful how you pose it. Let's simplify the discussion and look at a Young double slit experiment (the nature of the particle is irrelevant). A particle leaves a source and arrives at a screen by passing through one of two slits. Asking what the AMPLITUDE that it goes through one slit or the other is a well-defined question, but asking for the PROBABILITY it goes through one or the other is not, because all amplitudes (through both slits) contribute to the probability that the particle arrives at a given point on the screen. Framing the question in terms of the probability the particle goes through one slit or the other is in essence ignoring interference terms. More precisely, we can call A1 and A2 the amplitude that the particle goes through one slit or the other and arrives at a given point on the screen (each of these is a complex number). The total amplitude is A1+A2. The probability the particle arrives at the screen is
|A1+A2|^2, which is the sum of four terms, |A1|^2 + |A2|^2 + A1*A2 +
A2*A1. The first two are probabilities that the particle goes through one slit or the other; the last two are the interference terms. The danger is that if you take these probabilities too seriously, or misunderstand the meaning of the amplitudes, you might say the probability of going through the first slit is |A1|^2 and that of going through the second is |A2|^2, so the total probability must be the sum of the two -- an error.

The meaning of the individual probabilities is that if you were to cover up the first slit, say, then the probability of going through the second slit is |A2|^2 and vice-versa. But with both slits open you must add the amplitudes, not the probabilities.

Back to your question (ignoring the fact that you are talking about photons because as I said above the amplitude vs probability issue is independent of what kind of particle you are discussing). A particle goes from point a to b. You can consider a plane between a and b and ask what is the amplitude the particle passes through a point (c, say) on that plane. The amplitude is given by the product of the amplitude for going from a to c times the amplitude for going from c to b. But asking what is the probability that the particle passed through that point is dangerous (similar to the double-slit experiment), because the total probability of going from a to b is the square of the sum over all intermediate points of the amplitude, NOT the sum over all intermediate points of the probability. So you can certainly ask what is the probability it goes through c, but you cannot sum those probabilities over all points c and obtain the total probability.

So you can certainly evaluate the probability of going through c -- but be careful how you interpret the result! In terms of your question "Do all areas have equal probability, or does the probability diminish with distance from the line.": the answer is surely no -- the probability would decrease as a function of the distance between c and the straight line

From:	DT Froedge <dtfroedge@glasgow-ky.com>
Sent:	Friday, January 02, 2015 1:34 PM
To:	Mackenzie Richard
Subject:	QFT Gravitation

Hi Richard:

Hope you had a nice holiday and like the rest of us glad its all over.

Been working on some more off the wall stuff, and since it involves QFT, I was hoping you would look it over and offer your opinion. It's not very technical and an easy read.

The part that has interested me is connection between graviation, and the speed of light, and the fact that QFT can alter the local speed of light in a vacuum, and comes to play in by intense nonlinear photon beams.

The paper I'm working on is at:

   http://www.arxdtf.org/css/interfering.pdf

Particurlarly the section of gravitation.

The reference doccument on laser self focusing which cites alterations in the velocity of light as the result of QFT effects is:

   http://arxiv.org/pdf/hep-ph/0611133.pdf

If you dont have time or interest thats ok. Thanks,

DT

From:	DT Froedge <dtfroedge@glasgow-ky.com>
Sent:	Thursday, December 11, 2014 5:15 PM
To:	Mackenzie Richard
Subject:	Re: path integrals

Hi Richard:

I have a question on path integrals that I can't find discussed anywhere. It probably is discussed, but I just don't understand the math very well

I expect you already know the answer, or where I can look it up, so if you have time here it is:

Presuming a photon with a given frequency propagates along a line from point a to point b.

If at the middle of the distance along the line (or another point) a perpendicular plane is established.

Can one determine the relative probability per unit area, on that plane, of the photon propagating through that area, as a function of the distance from the line.

Do all areas have equal probability, or does the probability diminish with distance from the line.

Hope you and Anne are set for a good Christmas.

Thanks

DT

INDEX

and electric field scaling, 220-21
and electron models, 106, 235, 271, 278-84, 345
flaws of concept of charge, 174, 180-81
in particle solution to Klein-Gordon-Dirac equation, 291, 295-97, 300, 301, 308, 310, 314
probability density of Feynman photons creates effect of, 16, 55-67, 87-88, 121, 152, 153, 162-74, 180, 182, 214
and value of fine structure constant, 215-17
chirality, 59, 63-64
clockwork, 28, 34-36, 37, 38
coincidence probability, 132, 134, 138, 163, 167, 170, 196-97, 335
conjugate phase photons, 95, 97-99, 142, 149
conjugate wavefunctions, 170, 196-97, 329-33
consilience, nuclear-atomic, 19, 27, 31, 49
continuous fields, 149, 151, 175, 180, 195, 199, 202, 230
Couchot, F., 81n7, 118n27, 118n28, 119n30, 178n22, 226n29
Czarnecki, Andrzej, 223n8, 228n47, 250n22, 251n27

Dagoret-Campagne, S., 118n28
Davies, Paul, 194, 238
de Bianchi, Massimiliano Sassoli, 223n7, 226n34, 250n19, 250n21
deficit energy, 18, 27-32
Delbruck, Max, 110
Devreese, J. T., 311n4
diffraction, 57, 58
Dirac, P. A. M., 120n43, 144n8, 156, 176n7, 182, 311n5, 335, 337n10
Dirac equation, 14, 64-65, 93-102, 104, 106, 107, 141, 210, 280, 289-91, 293, 301, 305-6, 307
Dittrich, W., 111, 117n12, 119n36, 225n20
Djannati-Atai, A., 81n7, 118n27, 119n30, 178n22, 226n29
Doppler shifts, 25, 57, 58, 323, 330
Doran, Chris, 106, 107, 116n2, 228n52, 271, 287n3, 311n7
double slit, 197, 229, 246, 331-36, 346, 347
Dunne, G., 120n45

Einstein, Albert, 180, 329, 334, 336n7
electric force, 109, 114, 140, 146-47. See also charge; electromagnetism
electric field scaling, 209, 210, 220-22
flaws of electric field theory, 149, 151-52, 174
probability density of Feynman photons creates effect of, 55-56, 69-80, 83, 87, 128, 162-74, 198, 210-14

relation between gravitational and electrical forces, 14-16, 218
and vacuum polarization, 115
electromagnetism
electromagnetic fields, 101-2, 111, 156, 193-95, 209-10, 242, 245, 251-52, 291, 297, 307-9, 314-15
energy density, 242
gravitation as an electromagnetic phenomenon, 255, 266
particle interactions, 290-91, 308-9
waves, 195, 199, 243, 248
electron. See also electron, models of
calculation of mass of, 37, 183-87, 209, 232
first rest mass energy level of the universe, 11, 18, 24, 27, 29, 49, 53, 130, 131, 136, 169
ground state of nuclear particles, 27, 28, 49
internal binding created by probability density of Feynman photons, 9, 18-19, 29, 162-66, 196, 200
mechanics of, 131-37
structure of, 164-66, 275
electron, models of
and calculation of fine structure constant, 184, 232
and charge effects, 106, 235, 271, 278-84, 345
and Dirac equation, 93-102
Geometric Algebra model of, 271-87
summary of two-photon model, 106-15, 234-36
and twin paradox, 339-43
Electron Creation Radius, 2-3, 9-10, 17, 23-24, 29, 41, 53-54, 167
Electron Energy Radius, 134
energy hole, 18, 28, 30
EPR Paradox, 329, 334
Erber, T., 117n17, 225n25
Errede, S., 67n8, 144n10
escape energy, 20-21, 32
Euler, H., 103n7, 109, 116n4, 116n6, 116n7, 116n8, 116n9, 204, 224n16
Evans, J., 112, 120n41, 178n21, 228n53, 258, 263n10, 264n11, 265

Fermat's principle, 112, 205, 238, 255-58
Feynman, Richard, ix, 52n12, 80n1, 120n44, 128, 193, 194, 223n3, 248n2, 310n1, 326n6, 336n5, 347. See also Feynman action paths; probability density of Feynman photons
Feynman action paths. See also probability density of Feynman photons
and anomalous gyromagnetic ratio, 24, 54